D0442992

Praise for

Nanotechnology and Homeland Security

New Weapons for New Wars

"U.S. policy-makers and -shapers: READ THIS BOOK! Then get to work."

— Rocky Rawstern, Editor, *Nanotech-Now.com*

"Nanotechnology is an enabling technology that could have an impact on the world that dwarfs the Internet's impact on our daily lives. Mark and Dan Ratner have ably illustrated some of the roles that nanotechnology can play in our future, including how it could enhance national security, make soldiers more effective on the battlefield, or even help prevent attacks on our homeland. As a member of Congress who is active in advancing the development of nanotechnology, I encourage other policymakers, educators, and social visionaries to become cognizant of tomorrow's possibilities."

—U.S. Representative Mike Honda, Member, House of Representatives Committee on Science

"The nanotechnology revolution will have as large an effect on the earth and society as the information technology and biotech revolutions put together. Like any other science and technology, nanoscience and technology can be used for good or evil, and can have unintended consequences as well as intended. It's up to us to be informed world citizens and to use it for the betterment of mankind. The Ratners present a balanced view of some of the possibilities that nanotechnology has already created and may create in the future, without the hype. It is worth reading for the non-scientist just to become informed."

—Dr. Cherry A. Murray, Physical Sciences and Wireless Research Senior Vice President, Bell Labs, Lucent Technologies

"This book does an excellent job of introducing the field of nanotechnology to the layperson by showing its promise for security and defense—perhaps the most relevant sectors of society demanding advances that only nanotechnology can provide."

—Josh Wolfe, Managing Partner, Lux Capital, and Editor, *Forbes/Wolfe Nanotech Report*

"Nanotechnology and Homeland Security *provides the reader with the most important weapon of all—knowledge. It is as much a blow against ignorance and hype as it is a primer for how real nanotechnology should contribute to our future security. Mark and Dan Ratner confront the utopians and the alarmists by debunking both 'molecular assemblers' and 'gray goo.' This book is informative, thought-provoking, and very readable."*

—R. Stanley Williams, HP Senior Fellow, Hewlett-Packard Labs

"The book is a clear overview of the two subjects of nanotechnology and countering terrorism, but its special strength is the thoughtful way it weaves these two subjects together."

—R. Stephen Berry, Department of Chemistry and the James Franck Institute, The University of Chicago, Chicago, Illinois

"The Ratners hit the bull's eye again. They cut through the hype and illustrate the significant breadth, impact, and near-term realities of nanotechnology. This is a must-read for both the practitioner and the layman to understand why nanotechnology will revolutionize nearly all aspects of our daily lives."

—Matthew B. McCall, Venture Capitalist and Board Member, AtomWorks

"The Ratners provide a great overview of the potential of nanotechnology to solve key challenges that are on the top of everyone's mind. Once again, they do so in a manner that is accessible to everyone. A great read."

—Sean Murdock, Executive Director and Co-Founder, AtomWorks

"This book identifies many of the issues that need to be examined, and to be dealt with, if nanotechnology is to become a fully mature, fully productive asset to our nation and to the world."

—James Murday, Chief Scientist, Office of Naval Research

"The authors do an excellent job of using their expert knowledge to clearly communicate complex topics into a clear, well-organized examination of the impact of nanotechnology on national security."

—Lynn E. Foster, Jr., Squire, Sanders & Dempsey LLP, and Author of the seminal Nanotechnology Yellow Pages study

"This is the first example of an accessible book discussing the highly relevant field of nanotechnology and its applicability to homeland security in laymen's terms. The Ratners successfully cut through the hype surrounding the topic, while stimulating thoughts on many possibilities of the technology, especially in the defense and medical arenas. Mark Ratner is an internationally recognized expert in the field of nanotechnology with an in-depth knowledge in the area. He is respected by those in the academic and industrial research communities as a creative thinker with a long-standing track record of pioneering technical concepts for development of new materials."

—Dr. Susan Ermer, Lockheed Martin Advanced Technology Center, Palo Alto, California

Nanotechnology and Homeland Security

New Weapons for New Wars

Daniel Ratner

Mark A. Ratner

PRENTICE HALL
Professional Technical Reference
Upper Saddle River, NJ 07458
www.phptr.com

A CIP catalog record for this book can be obtained from
the Library of Congress.

Editorial/production supervision: *BooksCraft, Inc.*
Cover design director: *Jerry Votta*
Cover designer: *Anthony Gemmellaro*
Art director: *Gail Cocker-Bogusz*
Manufacturing buyer: *Maura Zaldivar*
Publisher: *Bernard Goodwin*
Editorial assistant: *Michelle Vincenti*
Marketing manager: *Dan DePasquale*
Publicity: *Heather Fox*
Publicity: *Newman Communications*
Full-service production manager: *Anne R. Garcia*

Publishing as Prentice Hall Professional Technical Reference
Upper Saddle River, New Jersey 07458

Prentice Hall books are widely used by corporations and government agencies for training, marketing, and resale.

Prentice Hall PTR offers excellent discounts on this book when ordered in quantity for bulk purchases or special sales. For more information, please contact:
U.S. Corporate and Government Sales
1-800-382-3419
corpsales@pearsontechgroup.com

For sales outside of the U.S., please contact:
International Sales
1-317-581-3793
international@pearsontechgroup.com

Printed in the United States of America
1st Printing

ISBN 0-13-145307-6

Pearson Education LTD.
Pearson Education Australia PTY, Limited
Pearson Education Singapore, Pte. Ltd.
Pearson Education North Asia Ltd.
Pearson Education Canada, Ltd.
Pearson Educación de Mexico, S.A. de C.V.
Pearson Education—Japan
Pearson Education Malaysia, Pte. Ltd.

For Betty, for Billie,
and for Pico

Contents

Foreword

Mark Ratner has been associated with nanostructures for quite some time; his work in molecular electronics started in the early 1970s. I was privileged to meet Mark in the late 80s because one of my Naval Research Laboratory employees, Dr. Forrest Carter, was championing the topic of molecular electronics. Our relationship expanded in the 90s when Northwestern won a Department of Defense (DoD) multidisciplinary university research initiative (MURI) project involving nanostructures. My association with Mark merited an early copy of the Ratners' first book, *Nanotechnology: A Gentle Introduction to the Next Big Idea.* It was so well written that I have used it as a model to emulate by the National Nanotechnology Coordination Office staff.

This present book addresses a very important topic. Nanotechnology will clearly be crucial to the nation's security—be it military (DoD), or homeland (coast guard, police, fire, emergency response, medical). Why do I say that? The book itself addresses the science/technology perspectives in Chapter 2 and then relates them to defense (Chapter 3), security (Chapter 4) and the economy (Chapter 5). Let me provide some additional background and historical perspective.

The DoD interest in nanoscience can be clearly identified as early as the late 1970s when its Ultrasubmicron Electronics

Research (USER) program was initiated. In the late 1970s the projection to size scales below micron was considered very adventuresome. Hence the prefix "ultrasub" implying a simple extrapolation from the comfortable "micron" size scale. Today we would have simply said nanoelectronics research (although the acronym USER is certainly more compelling than NER).

The USER program was followed in the 1980s by several programs that sought to develop and exploit the new tools—scanning tunneling microscopy and atomic force microscopy—that now provide our eyes and fingers for measuring and manipulating nanometer sized objects. The dream of Feynman discussed in Chapter 1 suddenly took on some reality. By the early 1990s major DoD programs were already being initiated to exploit nanostructures. For example, the Defense Advanced Research Projects Agency (DARPA) began the ULTRA (ultra fast, ultra dense) electronics program (a nanostructure program but still not called such); that program began to explore technology options that the electronics industries are now exploiting as they continue the miniaturization of electronics components into the nanometer size scale. As a second example, the Office of Naval Research began a program in nanostructured coatings that has led to improved mechanical components now (in 2003) onboard ships. It is estimated that the improved properties of those new coatings will lead to greater than $100M per year savings to the Navy. As the authors discuss in Chapter 3, nanotechnology impact on defense has begun, with the promise of much more to come.

The interest in nanotechnology wasn't exclusive to DoD scientists. Adm. David Jeremiah, in 1992 the four-star Vice Chairman of the Joint Chiefs of Staff, had nanoscience briefed to the Joint Requirements Oversight Council (JROC). This is the same Adm. Jeremiah cited in Chapters 3 and 6—clearly a prescient individual. In 1995, the Secretary of Defense, Dr.

William Perry, was briefed on the prospects of nanoscience and nanotechnology. By 1997, the importance of "nanotechnology" to the Department of Defense led to its designation as a "strategic research area." It wasn't until four years later that the National Nanotechnology Initiative was created, with the DoD an enthusiastic supporter.

Why all this interest? The book provides compelling details on some potential applications. Let me share some of the visions that have motivated my personal commitment.

Lower casualty rates are an important goal for full-fledged warfare, and even more so for military "police actions." Nanotechnology holds the promise to provide much greater information, connectivity, and **risk reduction to the warfighter**. The continued miniaturization of electronic devices will provide 100 times more memory (terabit of information in a cm^2). Processing speeds will increase to terahertz rates. Displays will be flexible and paper thin, if not replaced by direct write of information on the retina. Prolific unattended sensors and uninhabited, automated surveillance vehicles under personal control will be providing high data streams on the local situation. The marriage of semiconductors and biology will provide physiological monitors for alertness, chemical/biological agent threat, and casualty assessment. Nanofibers and nanoporous adsorbents will protect against CB threats while minimizing heat burdens and providing chameleon-like color adaptation for camouflage. The small size of the nanodevices will limit the volume, weight, and power burdens. The Army has created a University Affiliated Research Center at MIT to work toward this goal. Chapter 3 addresses these possibilities and more.

As discussed extensively in Chapters 3 and 4, nanotechnology will revolutionize our approaches to **Chemical, Biological, Radiological, and Explosive Detection and Protection** sensor suites with unprecedented sensitivity and selectivity;

protective clothing/masks with decontamination capacity rather than simply trapping the agents; and automated personal therapeutic systems that sense the presence of physiological trouble and take appropriate countermeasures. A workshop report on this topic can be found at *www.nano.gov/related_rpts.htm* . It isn't only human terrorist actions that need these countermeasures; nature has been a highly effective "terrorist" when it comes to biological threats.

The U.S. military annually inducts around 200,000 new people, 8 percent of its total personnel. These warfighters, many with only high school educations, must be trained in the use of increasingly sophisticated weaponry. This education/training challenge is exacerbated by the fact that military personnel are frequently in remote locations—on-board ship or based overseas—far from traditional education resources. Nanotechnology will enable the development of a **Virtual/Augmented Reality Training/Education/(Entertainment) system**—a highly effective teacher's aide. This affordable (about $100) system will customize its teaching modes (audio, visual, tactile) to the individual user for rapid learning. The K-12 education system in the U.S. has similar challenges, especially in teaching science and mathematics. This teacher's aide will likely be even more important to public education than to the military. Chapter 6 alludes to this impact.

Nanotechnology will enable us to replace the human in many dangerous warfighting missions. This is already true for surveillance platforms (e.g., Predator and Global Hawk unmanned aircraft). The **uninhabited combat vehicle** will have a nanotechnology-enabled artificial "brain" that can emulate a skilled warfighter in the performance of its combat missions. Removing the pilot will also result in a more effective aircraft (less weight without the human support paraphernalia and no human g-force constraints on turns). While the fighter aircraft will likely derive the greatest operational advantages, similar benefits will accrue to uninhabited tanks, submarines, and other military platforms. The same technology will provide automated transport for civilian applications. High value to a person like me who has come close to falling asleep at the wheel of my car many times (fortunately without harm—unless scaring myself witless counts as harm). And even higher value to those who would not be so lucky as to suffer only a scare.

As the authors have carefully stated (with proper caution to mitigate against a dot.com-like catastrophe), the potential for nanotechnology impact is truly exciting. The National Nanotechnology Initiative might be compared to a newly born baby. It is a lusty child, growing rapidly, and full of promise. Our challenge—as citizens—is to shape that child. This book identifies many of the issues that need to be examined, and to be dealt with, if nanotechnology is to become a fully mature, fully productive asset to our nation and to the world. Read the book (especially Chapter 6 on societal, ethical, and geopolitical issues), think, and participate in the child rearing.

James Murday
Chief Scientist
Office of Naval Research

Preface

Neither life, liberty, nor the pursuit of happiness can be assured without security. So assuring national security is one of the chief obligations of any form of government. The history of the United States has been almost blissfully free of invading armies—the oceans and the naval "wooden walls" of John Adams have kept this country strong and secure. Except for bandits in the southwest and a few German saboteurs in Maine, the United States has had no foreign invaders for more than a century.

During this time, homeland security issues were largely concerned with domestic, natural, and exotic threats. After a series of terrorist incidents, culminating in the events of September 11th, this freedom from invasion and sabotage can no longer be assured. After September 11th, many things changed. The Department of Homeland Security was established, some pretty draconian steps to assure security were taken, a large number of dollars were spent, and traditional civil liberties were endangered—all for the purpose of security enhancement.

The years since September 11th have been times of international and economic stress, but they have been largely free from any terrorist incidents in the United States. Therefore, it is probably time to examine more calmly how to "do" homeland security—how do we use the abilities of our people and our nation to guarantee the security without which civilization isn't civilization.

We face a wide variety of threats. These include the usual suspects: earthquakes, storms, tornadoes, fires, explosions, and spills of toxic materials and pollutants. But threats to homeland security now include terrorism and terrorist acts, perhaps involving biological, chemical, or nuclear agents (often called weapons of mass destruction). Other threats, including computer software security issues, economic threats, and international issues such as global warming and climate change, as well as the need for adequate and secure energy supplies, are all aspects of assuring security for the people. To do this, the very first requirement is clearly to have the right people in place, including police officers, firefighters, doctors, scientists, engineers, water purification experts, computer security professionals, and all the other specially trained folks who help assure the continuation of our domestic systems in the event of disaster, natural or man-made.

In addition to people, we also need programs. Duct tape and plastic sheets are fine in their way, but they are not substantive responses to real security threats. In addition to the traditional capabilities, we need new weapons for this war on insecurity. Some of these new weapons are offered by nanoscience and nanotechnology. These are new scientific and technological areas that deal with structures the size of large individual molecules. These structures, roughly 50,000 times smaller than the width of a human hair, cannot be seen or felt. Yet even though they are invisible, they are remarkably potent and useful. Because they exist at the ultimate design scale at which nature works, nanomaterials and nanotechnology can help in dealing with threats old and new.

This book is about nanotechnology and homeland security. It discusses first, in everyday terms, nanotechnology and nanoscience (Chapter 2). The bulk of the book has been devoted to some of the threats in the battle space (Chapter 3) and how to deal with them. It also considers some of the threats and protective mechanisms for homeland defense (Chapter 4).

But homeland security involves more than dealing just with toxins, tornadoes, and terrorism. It also involves providing economic, environmental, and educational security—these are discussed in Chapter 5, where their close relationship to nanotechnology is stressed. Finally, Chapter 6 deals with larger issues, including policy considerations and ethical ones. The photos and images that are throughout the book (and particularly in the color insert) are included to show that this is not just the technology of tomorrow—nanotechnology applications are already reality.

Science, technology, and society evolve together. Science never exists in a vacuum. While the traditional view is that science pursues the understanding of nature in a fashion totally separated from societal concerns, the facts are quite different: from Archimedes and Pythagoras through Leonardo da Vinci to Pasteur and Einstein, scientists have engaged in issues of importance in the society. In the contemporary world, this linkage has become tighter. It is driven both by economic issues of productivity and of support for scientific research, and by the fact that science and technology are uniquely capable of addressing huge societal needs from energy supply to food safety to new materials to diagnosis and conquest of disease.

This directing of science and of technology toward issues of importance in the society is an important trend. Society now both needs and expects technological advances. Lewis Branscomb has perceptively discussed these ideas under the name of "Jeffersonian Science." While this book does not directly discuss these aspects, they are the basis for the entire topic of *Nanotechnology and Homeland Security.*

This is not a scholarly book—there are very few citations, although a great deal of information can be found both in our previous book, *Nanotechnology: A Gentle Introduction to the Next Big Idea,* and on many websites, including *nanotechbook.com.* We thank our friends for remarks both helpful and challenging, and we thank the very talented, generous, and insightful people who contributed quotations used throughout the book. These remarks help point out some of the significant themes of the text, but the people who made them bear no responsibility whatever for the views of the authors, or their expressions in this book. We are also grateful to Stacy, Nancy, and Genevieve for superb editing and support; to Steve Berry for his thoughtful remarks; to Jim Murday for his excellent Foreword; and to Bernard and the staff at Prentice Hall for helping us complete this book. Mark is grateful to the NSF and the DoD for supporting his research on nanoscience.

No advanced training in science, engineering, law, or military affairs is needed to read this book. We've written it because both nanoscience and homeland security are very significant components of the current American and world scenes. We hope to enlist our readers' experience and interest. These are important issues, and we hope this book will help contribute to an understanding of them.

Introduction

"Eternal vigilance is the price of liberty." —*Wendell Phillips*

The Great Awakening

> *These are extraordinary times. And we face an extraordinary challenge. Our strength as well as our convictions have imposed upon this nation the role of leader in freedom's cause. No role in history could be more difficult or more important. We stand for freedom.... The great battleground for the defense and expansion of freedom today is the whole southern half of the globe—Asia, Latin America, Africa and the Middle East.... Yet [our opponents'] aggression is more often concealed than open. They have fired no missiles; and their troops are seldom seen. They send arms, agitators, aid, technicians and propaganda to every troubled area. But where fighting is required, it is usually done by others—by guerrillas striking at night, by assassins striking alone—assassins who have taken the lives of four thousand civil officers in the last twelve months in Vietnam alone—by subversives and saboteurs and insurrectionists, who in some cases control whole areas inside of independent nations.*

The reference to Vietnam is the only thing that differentiates this portion of John F. Kennedy's 1961 State of the Union address from one that might be heard today. Since the 1960s the nature of war and of national security has changed drastically. JFK sought a way to address a world where terrorism and unconventional warfare were the rule and not the exception.

America had—and still has—the strongest conventional and nuclear military in the world, but JFK realized that its ability to address these new threats was limited. He requested funds for a variety of programs to increase national security. Perhaps the most long lasting of these programs was the further establishment and expansion of the Special Forces, which have been a keystone of America's strategy in every conflict since.

Two years before Kennedy's address, in California, Richard Feynman, a Nobel Prize–winning physicist and one of the century's great scientific visionaries, was giving a very different speech:

> *I would like to describe a field in which little has been done, but in which an enormous amount can be done in principle. This field is not quite the same as the others in that it will not tell us much of fundamental physics (in the sense of, 'What are the strange particles?') but it is more like solid-state physics in the sense that it might tell us much of great interest about the strange phenomena that occur in complex situations. Furthermore, a point that is most important is that it would have an enormous number of technical applications. What I want to talk about is the problem of manipulating and controlling things on a small scale.*
>
> *As soon as I mention this, people tell me about miniaturization, and how far it has progressed today. They tell me about electric motors that are the size of the nail on your small finger. And there is a device on the market, they tell me, by which you can write the Lord's Prayer on the head of a pin. But that's nothing; that's the most primitive, halting step in the direction I intend to discuss. It is a staggeringly small world that is below. In the year 2000, when they look back at this age, they will wonder why it was not until the year 1960 that anybody began seriously to move in this direction.*

Feynman went on to lay out his vision for this entirely new field. He described computers that could adapt to changing circumstances, primitive artificial intelligence, and data storage that could put the contents of every book in the California

Institute of Technology library onto a single library card. Moreover, he spoke of dozens of applications for this new field that hinged on the ability to manipulate matter at its ultimate scale, atom-by-atom and molecule-by-molecule. This was in the days when vacuum tubes were common, and computers were the size of a whole room yet less powerful than a modern cellular phone.

The challenge was not taken up immediately. Instead, miniaturization continued to take its "primitive, halting steps" forward toward the information technology revolution in the 1990s. It wasn't until close to the year 2000 (as Feynman foretold with uncanny accuracy) that new tools to see and work at the level of atoms and molecules became widely available. At around the same time, it became clear that microchips could not continue to be improved by evolving current techniques. Feynman's science of the very small suddenly became both possible and necessary. It became known as nanoscience, and its applications became known as nanotechnology.

On September 11, 2001, America became the victim of the most violent terrorist attack it had yet endured—an attack carried out by fewer than 20 men using knives and box cutters killed more than 3,000 people in New York, Pennsylvania, and Washington, D.C. The attacks broke the nation's heart and forced its citizens to reconsider many of JFK's observations. How can a country protect itself against forces that, while possibly state-sponsored, are not the troops of a sovereign nation? How do we protect our children, airplanes, roads, schools, buildings, and infrastructure against the threat of terrorism?

Terrorism was not new to America. The World Trade Center had been bombed before, Pan Am flight 103 was hijacked, Al Capone bombed businesses that would not pay protection money, the Unabomber sent explosives to leading scientists, and Timothy McVeigh attacked the Oklahoma City Federal Building. Nonetheless, an attack of the size and scope of 9/11—carried out just as America's economy was entering a recession after its euphoric "tech bubble" years—resulted in a great awakening to the fact that all was not right in the world and that we were not, even on our home soil, anywhere near so safe and omnipotent as we had thought.

The 2003 war in Iraq has brought these points into even sharper focus. If terrorists or countries sponsoring terror also possess chemical and biological weapons, what can we do to defend our troops and our cities? If our only mode of response to a terrorist attack is to invade an entire nation, how do we protect our people at home from revenge attacks, and how do we assure the safety of innocent civilians in the countries we invade? How do we work quickly and cleanly to avoid having matters spiral out of control the way they did when a terrorist's bullet started World War I? And finally, how do we find and stop the individuals, such as Osama bin Laden or Saddam Hussein, who are most responsible?

The government's initial responses to these problems—the formation of a unified Department of Homeland Security; a new Transportation Security Administration; increased funding for local police, fire departments, and other "first responders"; and a renewed emphasis on intelligence gathering and interagency coordination—were important but not sufficient. Faced with the grim possibilities of chemical or biological attacks and with the increased threats of terrorism brought about by the wars in Iraq and Afghanistan, some of these multibillion dollar agencies did little more than increase the national threat advisory level, recommend limited inoculation for some common bioagents such as anthrax and smallpox, and suggest that citizens purchase duct tape and plastic sheeting to create a protected room at home.

None of these precautions were terribly helpful or workable. Few civilians understood the obscure color-coded system of the national threat advisory level, and they were given no practical recommendations for heightening their security when the level was increased, other than to curtail travel and other unnecessary activities, such as shopping—something which did not help a depressed economy. Sealing rooms with plastic sheeting and duct tape would do little or no good in the case of attack, since any room that will seal out toxins will also seal out breathable air. Inoculation regimens for certain biological agents made a bit more sense, but there are so many different biotoxins and biotoxin strains that inoculating against all of them is impossible. Finally, renewed intelligence efforts have certainly helped in catching terrorists, but they also resulted in policies such as the USA Patriot Act and the proposed Total Information Awareness Act that may destroy the very freedoms and liberties we are trying to protect. As Benjamin Franklin, scientist and politician, remarked, "They that can give up essential liberty to obtain a little temporary safety deserve neither liberty nor safety."

In the battlefield, the situation is little better than in the homeland scenario we have just discussed. When facing the threat of chemical or biological attack, soldiers had either to wear thick chemical defense suits, which were similar in weight and comfort to SCUBA suits and made desert fighting next to impossible, or to take their chances in the face of these serious threats.

While the search for long-term solutions to these problems continues, what America and all the other nations of the free world need is an arsenal of "silver bullet" technologies that address these very specific needs. The military needs lightweight uniforms that allow troops to fight in the desert while protecting them from bullets, shrapnel, chemical gases, and biological toxins. It also needs better smart munitions; improved stealth features; tougher, lighter vehicles and battle gear; more unmanned combat capabilities; and better battlefield information systems that emphasize smaller units well inside hostile territory. UN inspectors need to be able to find and trace weapons of mass destruction quickly and efficiently. Water treatment facilities, mail sorting rooms, and transportation centers need sensors that can find explosives and other dangerous cargoes immediately, in small quantities, and without error. First responders need substances that can neutralize chemical and biological warfare agents. Doctors need treatments to help those who have been exposed. Industry needs more secure and robust communications and computer systems. Law enforcement agencies need better capability to identify, intercept, and decode information from potentially hostile groups. And everyone needs new economic drivers and alternative sources of clean, cheap energy that will solve environmental problems and reduce our dependence on the foreign oil that so often comes from troubled parts of the world.

All of these things and many, many more are being made possible by nanoscience and nanotechnology, converging the visions of Feynman and JFK. These sciences have prompted Clifford Lau, director of the Office of Basic Research at the U.S. Department of Defense, to say that nanotechnology will eventually alter warfare more than the invention of gunpowder. While policy and the people in all the uniformed services are crucial to security, without the tools made possible by nanotechnology, the world may not be much safer than it was before September 11th. With the first wave of intense fighting in Iraq and Afghanistan over, the urgency of these issues may seem to have diminished. It has not. Complacency in the wake of victory is among the most dangerous mistakes to make in war. Security is a long-term problem, and though it may not be the sole obsession of most citizens (as it has been during most periods of human history), it remains a great concern and the first duty of government.

The Case for Nanotechnology

Nanotechnology has immediate applications for defense and security, and they are the focus of this book, but it also has broader possibilities that will eventually touch every aspect of our lives. Many of the applications discussed here will have a profound impact on how we deal with disease outbreaks such as AIDS or SARS. Energy production, environmental repair and remediation, and advanced computing are also so-called dual-use applications, where industry and the military can pool resources for mutual benefit. There is also a host of completely nonmilitary applications—nanotechnology is already used for making better clothing, sports equipment, and treatments for diseases from diabetes to cancer—but much of that is beyond the scope of this book. Still, the applications abound. As Shimon Peres, former prime minister of Israel, observed, "That which has been achieved by the atomic bomb in the field of military strategy will be accomplished in the future by nanotechnology in the field of civil potential."

Unfortunately, technology is often a double-edged sword. The cellular phones and airplanes that are so much a part of everyday life were also necessary elements to the terrorists' plot on 9/11. Nanotechnology could be the same. Never before has a single branch of technology promised to be so transforming or powerful, nor has a technological revolution happened so quickly. It is of enormous importance that we examine both the promise and perils of nanotechnology and that we embark on this task quickly, with our full attention.

With the first nanotechnology products already appearing on store shelves and on the battlefield, we cannot afford to wait.

There has been a great deal of hype about nanotechnology. Enthusiasts say that it will solve all the world's problems, from ending hunger to achieving eternal life. Critics cite concerns ranging from regulatory and intellectual property issues to the idea of "gray goo," usually meaning nanotechnology-enabled, self-aware, self-replicating machines that would ultimately take over the world. Different views of nanotechnology inform movies, TV, and science fiction from *Star Trek* and the *X-Files* to Michael Crichton's thriller *Prey*. Almost none of these portray what nanotechnology is really about. Nanoscience is not the science of creating invisible robots to assemble kitchen sinks out of the air or to invade a human body. It is the science of utilizing the unique properties of the design scale of nature to make materials and perform functions that could not be made or performed any other way. While it might be exaggerating to say that nanobots (those tiny, invisible robots) are absolutely impossible, there is a compelling body of evidence showing that their near-term development is extraordinarily unlikely.

Criticism of nanotechnology also takes another form. Many observers point out that while there has been a great deal of hype and a reasonable amount of investment in nanoscience and nanotechnology (governments and private firms worldwide spent some $4 billion on these fields in 2002, though less than half of the total was spent in the U.S.), there are few nanotechnology products to justify the excitement. This is unfair and a bit premature. While $1 billion may sound like a great deal of money, it is widely distributed for a host of new applications, and it is less than the military often spends to develop a single new weapons system. (Ballistic missile defense, for example, consumes some $8 billion per year with far fewer results.) It is less than 3% of the cost of the war in Iraq. And it is smaller than the internal R&D budgets of some individual pharmaceutical firms. In military development terms, nanotechnology still has a very small budget and a small program, but it is growing rapidly.

There are other flaws in the argument that nanotechnology is more hype than substance. Most of the dollars spent to date have been on fundamental research, and it may take several years to develop the results into useful products. Also, some heavy-hitting, nanotechnology-based products are already on the market: zeolite technology for oil refining currently saves the U.S. some 400 million barrels of oil a year, and GMR (giant magneto-resistance) technology is what makes multi-gigabyte computer hard drives possible. Since not all nanotechnology products have labels saying "Nano Inside," it is sometimes difficult for consumers to know which products are nano-enabled and which are not. However, with the U.S. defense budget above $400 billion per year, and the combined market impact of global defense and energy at more than $4 trillion, if nanotechnology achieves its expected impact over the next 10 years in these sectors alone, the National Science Foundation's estimate that it will be a $1 trillion industry by 2015 could, if anything, be rather low.

This book examines some of the more immediate and key areas of nanoscience and nanotechnology as they apply to homeland security and to national defense. It emphasizes the capabilities of these new fields and makes some recommendations for integrating them into policy while avoiding the perils they pose. There is also a significant section on the ethical aspects of nanotechnology. The book is intended to be almost entirely nontechnical except to the extent that is necessary to explain the basic ideas. For a more general description of nanotechnology, more detail on how it works, or more analysis of the business of nanotechnology, we encourage you to read our previous book, *Nanotechnology: A Gentle Introduction to the Next Big Idea* (Prentice Hall, 2002).

Nanotechnology: What and Why?

David Swain
Executive Vice President and Chief Technology Officer
Boeing Corp

If it weren't for breakthroughs in structures, materials, propulsion, and computers, we wouldn't be flying millions of people around the world every day or sending people and satellites into space.

If it weren't for advancements in stealth, precision guidance and C4ISR(Command, Control, Communications, Computer, Intelligence, Surveillance and Reconnaisance) technology, we wouldn't be resolving military conflicts as quickly or effectively and survivably as we are today. It has taken 100 years since Kitty Hawk to get to where we are today, and I think it will take us about 25 more years to make our next big leap in progress.

Nanoscale science and engineering most likely will produce the strategic technology breakthroughs of tomorrow. Our ability to work at the molecular level, atom by atom, to create something new, something we can manufacture from the "bottom up," opens up huge vistas for many of us. Breakthroughs could bring us nano-structured metals; ceramics and polymers at exact shapes without machining; nano-coatings for cutting tools and electronic, chemical, and structural applications; nano-instrumentation for micro-spacecraft avionics; nano-structured sensors and nano-electronics for embedded health management systems in aircraft structures and systems; and thermal barrier and wear-resistant nano-structured coatings. There are huge possibilities for nanotechnology applications. This technology may be the key that turns the dream of space exploration into reality.

Of course, all of this assumes a level of maturity in nanotechnology that remains to be achieved. Meanwhile, we have been watching the progress in this field and even using some of the concepts and early results for conducting some of our own early-stage projects.

In a program called "Structural Amorphous Metals," for instance, Boeing is helping develop a process for controlling the nanostructure of aluminum to make it as strong as titanium while remaining light as aluminum—to improve vehicle bodies and structures.

In another project, we are using nanomaterials to improve the peel and delamination resistance properties of composites in order to eliminate the need for fasteners.

In our "Optical Beam Steering" program, we are applying nanoscale optical materials to produce ultra-small laser communications devices. In addition to being able to operate in military environments, these devices are also much lighter, more energy efficient, cooler, more powerful and more reliable than previous devices.

Reaping the benefits from this science is a long-term journey that will be challenging for all those involved. And, while some of the near-term applications are encouraging, it is the long-range possibilities that have the potential to change our world at the most fundamental level and lead to an unparalleled economic revolution. Because we are at just the beginning of this potential revolution, I reiterate that a high commitment to this journey will be a critical factor in successfully achieving the potential of nanotechnology. To this end, I encourage us all to take the necessary steps to make the dreams of nanotechnology come true. It is up to us to make the 21st Century the nano-age!

This book is about international defense, homeland security, and nanotechnology. To most people, the first two topics are pretty straightforward, and the last seems a bit confusing—it is the stuff of science fiction, of *Star Wars* and *Spiderman.* This chapter introduces what nanotechnology is and why it is important. It focuses on the overall properties of nano and serves as the background for the security, economic, and policy chapters in the remainder of the book.

What is nanotechnology? By definition, it is the application of nanoscience to useful devices. Nanoscience is in turn the science that deals with objects with at least one dimension between one and one hundred nanometers in length, a size range called the nanoscale. A nanometer is one one-billionth of a meter, which is pretty close to one one-billionth of a yard. For comparison, a human hair is approximately 50,000 nanometers across, and a nanometer is as much smaller than a football as a football is smaller than distance from the earth to the moon. Anything small enough to be measured in nanometers is much too small to be seen with the naked eye.

So if nanoscience is the science of dealing with objects whose size is almost inconceivably small, why does it get so much hype, and why is it so important for national defense and homeland security? The first reason is that nanoscale objects are not just small, they are a special kind of small. Individual atoms are around one-fifth of a nanometer. The size of almost all molecules from alcohol to sugar to caffeine lies within the nanoscale, because it is the smallest level at which functional matter can exist—anything smaller is just a minute speck of vapor. Material designed at the nanoscale can therefore be designed with molecular precision. This means that, through nanotechnology, we can make materials whose amazing properties can be defined in absolute terms: "This is not only the strongest material ever made, this is the strongest material it will ever be possible to make."

There is another aspect of the nanoscale that makes it important. It is the scale at which the quirky quantum mechanical properties of matter and its more familiar mechanical properties (such as hardness, temperature, and melting point) meet. At the nanoscale it is possible to take advantage of both sets of properties, and this allows us to do things that simply cannot be done any other way. This duality is essential to the basic processes of life, which is why nature builds at the nanoscale. Most fundamental biological structures including DNA, proteins, and enzymes do their jobs at the nanoscale (cells are much larger), working molecule by molecule to build the macroscopic structures we call leaves, tadpoles, weevils, and humans.

The properties of smallness, then, are what make nanoscience and nanotechnology so important. By creating structures whose size is the same as individual molecules or collections of molecules, we can create devices with new and unique properties that solve many of the most intractable problems. Nanofabrication techniques let us make bulk quantities of materials designed at the nanoscale and bring these properties into the macroworld. The potential applications of nanotechnology run the length and breadth of society and of industry, but the ones with which we will be most concerned in this book are those involving security.

What Nanotechnology Is Not

There are a few common misconceptions about what nanotechnology is. For the most part, these misconceptions arise from the portrayal of nanotechnology in science fiction, and they have become so fixed in the public attention that they distract from the genuine promise of the nanoscale. The first of these misconceptions is the concept of a molecular assembler. The other is the concept of "gray goo."

Let's start by examining molecular assemblers. This idea is derived loosely from the writing of Eric Drexler, founder of the Foresight Institute and coiner of the term "nanotechnology." The idea is that a device could be created to place individual atoms together to form any arbitrary nanostructure desired. As the device assembled these structures, they could in turn be placed together to make macroscopic amounts of any given material or manufactured item from perfect rubies to kitchen sinks to armored tanks. If they existed, these assemblers could be carried along in battle to repair damage and all repairs would be absolutely as good as new. On the surface, the idea sounds futuristic but plausible, given nanotechnology's ability to work at the molecular scale. Clearly the benefits of such a device would be enormous. A factory of molecular assemblers could make any item without flaws, and it could go into action on the battlefield to do everything from healing wounds to disassembling enemies' equipment.

After a bit of closer scrutiny, though, the cracks in this argument begin to show. There are a number of compelling reasons why molecular assemblers are either impossible or are at best in our distant future, and it's worth looking at a few of them in order to read sci-fi without nightmares.

One way to understand the challenges of molecular assemblers is by thinking about the Lego robot example. Suppose that you want to build a robot out of Legos whose job it is to build other things out of Legos. The only way for your robot to reliably move a Lego piece would be to attach to it in the traditional Lego brick fashion, but then it could not easily let go. (In essence, it has sticky fingers—no fair reaching in to remove the Lego from the manipulator.) Your robot would also have trouble placing additional parts in tight places, because the bricks making it up are too big (for, though we do not wish to be unkind, it has fat fingers). You have no command and control device to tell your robot where to put the next Lego (you can't just connect a computer, because even a simple circuit is a million times bigger than your robot), and you have no power source to make your robot go.

All in all, then, while your Lego robot is probably very cute, it is hopelessly unsuitable for building more things out of Legos, and all of its problems are analogous to those faced at the molecular scale. Sticky fingers that cannot release parts are the result of inconvenient chemical interactions. Fat fingers that cannot operate in tight spaces, the power supply issue, and a lack of control circuitry arise because the nanoscale is the ultimate scale of miniaturization and can't accommodate macroscale devices like batteries. Even Nature's best assembler, the cell, can only accommodate all this machinery by being much larger than the nanoscale.

If one Lego robot cannot do the job, might we get somewhere with an army of them? No. Without a power source, communications, or control circuitry, these robots could not work in parallel. There would be no way to coordinate action analogously at the nanoscale. This means that a single assembler would need to act to build up your materials atom by atom. Even assembling at the rate of millions of atoms per second, it could take billions of years to generate even a few grams of target material.

Finally, even beyond these problems, there is the issue of molecular stability. Molecules resemble card houses in some ways. Once they are completely fitted together, they are stable, but there are many stages during their construction at which they will just fall over if not supported. There are design rules arising from quantum mechanics—only some card house structures are stable. For example, carbon atoms can bond to one, two, three, or four other carbons, but not to five or more. Working alone, a single manipulator placing atoms in a structure could not provide this support. This could result in any number of problems; to pick a simple one, a robot trying to make water molecules (H_2O) from supplies of hydrogen and oxygen by mechanical assembly (using manipulator arms rather than allowing chemical reactions to happen spontaneously) might instead produce little clouds of hydrogen gas (H_2) that could go wafting off before the final oxygen could be added.

Making particular nanostructures is of course totally practical. As we discussed extensively in *Nanotechnology: A Gentle Introduction to the Next Big Idea*, both synthesis of molecules and fabrication of nanostructures are quite advanced areas of science and technology. Almost any pharmaceutical plant makes nanosized molecules, from Tylenol to Vancomycin to Viagra. The process includes chemical reactions and physical separations like crystallization, with no hint of nanoscale controlled robots.

Some of the newest fabrication methods can indeed prepare a wide variety of individual nanostructures. For a beautiful example, Don Eigler's group has made the IBM initials in letters a few atoms tall. But the tools used are macroscale probe tips, and the structures are stabilized by sitting on microscopic metal surfaces. Even then, manufacturing nanostructures this way a single machine can yield only .000000000000001 ounces of material per year.

So, while the idea of molecular assemblers is great, practical considerations get in the way.[1] We will have to find other means of making self-healing materials for use in battle and this book will suggest a few of them.

The other common misconception about nanotechnology is the nightmare "gray goo" scenario described by people including Bill Joy at Sun Microsystems, Prince Charles (who

1. For a very different viewpoint, compare the Web site *www.imm.org*.

probably got his ideas by divine right), Chris Carter, and Michael Crichton. The idea here is that little nanomachines (described as gray goo) could invade the blood stream of people or the computer systems of starships, communicate with one another, self-replicate, and possibly colonize our bodies, control our minds, or simply consume or destroy us.

While the gray goo scenario and its derivatives are certainly frightening ones (and great bases for good science fiction novels), they are emphatically not what nanotechnology and nanoscience are about. Self-replication is one of the secrets of life, but we know enough about it to realize that simple nanostructures, of the sort that nanotechnology will try to make, can't do it. Cognitive intelligence of the kind envisioned in this scenario is also impossible at the nanoscale. Human-level intelligence requires extraordinarily complex machinery and intricate communication channels. The average human brain weighs about three pounds. It is certainly not nanosized. Even the functional components of human brains, nerve cells, are much larger than the nanoscale and need to be fed by an equally complex system of blood vessels and protected by bones and membranes. In short, the gray goo scenario is far-fetched for many of the same reasons that molecular assemblers are. Functions this complex simply don't work at a scale this small, and even if nanobots were the size of bacteria or viruses, we have the human immune system and a host of other technologies for defending ourselves against them.

Major Branches of Research

The reason we spent so much time examining the feasibility of molecular assemblers is to demonstrate that nanotechnology's primary importance does not lie in finding new ways to make existing products, but rather in the opportunities it offers to make entirely new things that can only be created at the nanoscale. The scope of such creations is vast. In the context of defense and homeland security we are most interested in seven major areas: materials, sensors, biomedical nanostructures, energy, electronics, optics, and fabrication. Each of these major areas will be examined on the following pages.

Materials

Materials science, which deals with the properties and structure of materials, is the youngest branch of engineering. Materials designed at the nanoscale have remarkable properties because their exact molecular or atomic structures can be precisely tweaked. A natural example of this is the difference between diamond and graphite dust. Both are made up of 100% carbon atoms, and the only difference between the diamond in an earring and the black stuff in a pencil is molecular structure. Nanotechnology is concerned with applying this principle to man-made objects. Examples of these nanoscale-designed materials include ultrathin molecular coatings that turn ordinary cotton cloth into water-repellent, stain-resistant fabric and ultrahard coatings that make materials resistant to scratching and abrasion. Other nanomaterials include zeolite molecular sieves, whose nanoscale pores and channels can be used for applications such as petroleum refining, oxygen separation from air and water softening, or piezoelectric materials, which can change their dimensions when electrical current is passed through them.

Perhaps the most remarkable example of a nanomaterial is a new member of the carbon family called a nanotube, shown in Figure 2.1. Nanotubes were discovered in the 1990s and they come in several varieties, but the most researched are single-walled nanotubes that look like nano-

Figure 2.1 Image of a single-walled carbon nanotube, viewed from the inside. Each of the bonds is roughly 0.14 nanometers long. Courtesy of Chris Ewels, *www.ewels.info*.

scale soda straws. Like diamond and graphite, nanotubes are made out of only the element carbon. Their diameters can range from less than one nanometer to near 20 nanometers, and they can be best envisioned by imagining chicken wire, wrapped up onto itself and then welded together. Each three-wire crossing point within the chicken wire is analogous to a carbon atom in a nanotube, but of course the nanotube roll is ten million times smaller. Nanotubes are the strongest, most conductive, and stiffest materials ever made (over 60 times stronger than steel by some estimations), and they are designed at the nanoscale. Some scientists have estimated that a nanotube cable the size of a human hair could suspend a locomotive or reach from the Earth to the moon while bearing its own weight.

Materials designed at the nanoscale are often called smart materials since they are custom designed for a particular purpose. They can also be made to react to outside stimuli, as in the case of self-tinting window glass which darkens in direct sunlight. These are called dynamic smart materials and will be key to many military applications.

Sensors

A sensor is any device that gives a recognizable signal in response to the condition or thing it is designed to detect. This signal can be any discernable change in the properties of the sensor such as the color change of litmus paper in the presence of acid, the shriek of a burglar alarm set off by an intruder, or the vibration of a cell phone sensing a call. In nature, noses, eyes, and other organs act as sensors.

Nanoscale sensors are generally designed to form a weak chemical bond to the substance of whatever is to be sensed, and then to change their properties in response (that might be a color change or a change in conductivity, fluorescence, or weight). For sensing biological species such as toxic organisms or bioterrorism agents, sensors usually work by specifically binding to the target DNA of the given biological agent.

Sensors could be used to detect such substances as oxygen, anthrax toxin, carbon monoxide, or nerve agents. Development of sensors is one of the major applications of nanotechnology in global security.

Biomedical Nanostructures

Biomedical nanostructures are designed to interact with particular molecular or larger scale structures within a biological organism, particularly within people. Simple biomedical nanostructures are used in drug delivery to encapsulate small amounts of a drug, thus speeding its dissolution within the body and its availability at a local site where remediation of pain or disease is necessary. Magnetic nanomaterials can also be used to aid in drug delivery, for they can be attached to a drug molecule and then directed externally, by magnetic fields, to different parts of the body. Applications of biomedical nanostructures in the treatment of disease include adhesive materials for skin grafts or bandages, oxygen barrier structures for burn therapy, and tiny metal or semiconductor tags that can be attached to particular proteins or medium-size molecules and used, with microscopes, to observe the actual biological function or structure of these molecules. Such bionanostructures will be crucial in assuring effective homeland and military defense.

Energy

In addition to homeland defense, global climate change and the energy supply are major issues in national and international security. Nanostructures play key roles in scenarios for reducing climate change through reducing the burning of carbon-containing materials, and for making the energy supply entirely independent of hydrocarbons such as petroleum. Particular nanostructure applications in energy include the use of titania nanoparticles (titania is the white pigment in house paint) to capture electricity from the sun. They include fully integrated nanoscale fuel cell structures for computer backup power and portable energy, ultrathin separation layers for advanced batteries, and transparent electrodes for energy efficient lighting using so-called light emitting diode structures. Finally they include nanoporous carbon electrodes for advanced, high energy, and high power battery structures.

Electronics

Modern computers are inexpensive and effective because their manufacture actually involves nanoscale structures. Ever since the transistor, which is still the fundamental element of a computer, was developed at Bell Laboratories in the 1940s, it has been getting smaller. The length of the current transistors is roughly 100 nanometers, and therefore these transistors exist at the nanoscale. Transistor materials include silicon, silicon nitride, silicon dioxide, gallium arsenide, and other exotic semiconductors. But despite existing at the nanoscale, semiconductor electronics do not yet take advantage of any of the unique properties of the nanoscale. They simply represent the continued miniaturization of components. For this reason, modern semiconductor manufacture is not considered by many to be true nanotechnology. Unlike nanotechnology, it is a technology that is reaching maturity and approaching its fundamental limits. Nanotechnology can open the doors to new possibilities.

Some nanotechnological approaches to the problem of continuing to shrink circuits and make them more efficient change the rules at a very basic level. For example, scientists are now able to build circuits with individual molecules as circuit elements. Indeed, it was Richard Feynman's speech in 1959 entitled "There's Plenty of Room at the Bottom" (quoted at the start of Chapter 1) that began to turn scientific thought toward nanoscience. In that speech, Feynman talked specifically about electronics applications of nanostructures in computing and in memory.

Not all electronic applications of nanostructures are involved in information technology, though. Nanometer-thick conducting wires can also be used to dissipate static electricity. In addition, both light emitting diode structures and candidates for all-optical computing utilize nanoscale structures for their function. Given the crucial importance of information and electronics in defense at all levels, nanoelectronics will be a featured component of future security scenarios.

Optics

Optics is the study of light, but by nanoscale optics we generally mean the interaction of light with nanostructures. Applications of nano-optics include control of color by control of nanoparticle size (this is called the quantum size effect, and it is one of the most striking results in nanoscale science: pure gold can range in color from green to dark red, depending on the size of the nanoparticles). Materials with controlled fluorescence can also be designed at the nanoscale, and once again their properties, including their colors, can be tuned. Some of the other applications mentioned above, including light emitting diode structures and photovoltaic (conversion of sunlight to electricity) devices, are also applications of nanostructures in the general areas of optics. Finally, active camouflage is a very promising nanoscale-based soldier protection scheme, and it is described in the next chapter.

Fabrication

Fabrication is perhaps not so much an application as a tool, but it is of major importance: if we are to understand and study nanostructures, we need to be able to make them. For example, NanoInk, a Chicago-based startup, uses nanoscale structures as pens to write very narrow (a few nanometers thick) inklike lines of specific molecules on surfaces. This is called dip-pen nanolithography. It is also possible to employ spheres of nanoscale dimension laid out on a surface to act like a mask for spray painting, resulting in a pattern of hollow triangles (made of molecules that behave like the paint) of nanoscale dimension isolated on the surface. This is called nanosphere liftoff lithography. Continuing the analogy between macroscale and nanoscale patterning, we can make stamp pads of a soft plastic material called PDMS, that can be used to stamp out arbitrary patterns with molecular "inks" on appropriate surfaces.

Since unlike charges attract, we can use the attraction between positive and negative charges to make layers only one molecule thick. This is done by first laying down a layer of positively charged molecules, then a second layer of negatively charged molecules, then the third layer of the positively charged molecules again, and so on. This fabrication method can be used to design very thick (macroscopic) structures that are homogeneous on the macroscale and layered at the nanoscale.

Although these capabilities have been developed in academic labs by the groups of Chad Mirkin, Rick van Duyne, George Whitesides, and Mike Rubner, respectively, they are now being pursued actively by startups, often with defense applications in mind.

As already indicated, nearly all of the major devices used in modern computational structures, such as desktop, laptop, or workstation computers, are based on some rather simple structures. One reason it costs so many billions of dollars to build a fabrication plant for computers is that as the components shrink to nanoscale dimension, unusual forces such as the capillary force (which also makes wet sand castles much stronger than dry ones) and other molecular binding interactions can be the major directors of structure. Learning to use the resulting structures to do computations is a major challenge. Feynman first suggested that if we are to continue to make computers smaller and smaller, we will eventually have to stop making them by carving up large slabs of silicon and start building them the way nature builds us—by assembly of molecules starting from the bottom and getting bigger. This kind of bottom-up technology will characterize much of the nanotechnological world.

Why Nanotechnology?

Nanoscience/nanotechnology is important and exciting not simply because of its small dimensions. "Small is beautiful" is a cute phrase, but unless there are real advantages in using nanoscale structures, the trouble isn't worth it. It is important and exciting because there are unique capabilities offered by working at the nanoscale. Perhaps the most important of these is the ability to control the interaction of nano devices with other systems: with molecules to be sensed, with sunlight to be converted to electricity, with pollutants or toxins to be kept out, with viruses or bacteria to be destroyed, with injured or diseased biological structures to be repaired, or with sound to be amplified in replacing hearing loss. In each of these cases and hundreds more, working at the nanoscale has unique capabilities.

These unique capabilities, as we have said before, come from the fact that in nature, certain properties occur only at the nanoscale. These include so-called molecular recognition—the ability of molecules to bind specifically to other molecules in unique and understandable ways. This capability is what makes biology work, and therefore if we are going to interact in a positive way with biological systems, repairing or replacing or augmenting or reducing difficulties, the agents that we use must be nanosized. Similarly, if we are to control where molecules go (everything from wicking away perspiration to controlling frost heaves and cracking on road surfaces), the control must be at the nanoscale. For example, the flickering metal nanodots that are used to label biological molecules with light emitting properties only work when those dots are nanosized. The photovoltaic capture of sunlight using titania cannot be done with bulk particles, but only with nanosized grains.

The next question facing nanotechnology is not "Can these nanotasks be done?" but rather "Can they be done efficiently, cheaply, and in an integrated fashion?" These are challenging issues. Although the nanoscience revolution is in full swing, the issues of reliability, cost capability, manufacturability, utility, compatibility, and recyclability have not yet been extensively investigated. Still, the elegance of the small is striking: the enormous complications and high costs associated with making something as complex as a jetliner, an automobile, or a milling machine cannot occur at the nanoscale simply because the structures are too small. As we learn how to build and manipulate nanostructures better, nanotechnology should become inexpensive and widespread, providing unique advantages in everything from cosmetics to security.

Although individual nanostructures are too small to see with the eye or feel with the finger, extended nanostructures are visible, tangible, and important. Manufacturing capabilities for these structures can sometimes be developed very quickly. For example, two startup companies in the Chicago area have the capability to manufacture extended nanostructures for materials applications. One, Advanced Diamond

Technologies, uses technology originally developed at Argonne National Laboratory to coat structures with nanoscale diamond. This diamond coating is, like diamond itself, very hard, chemically resistant, and atmospherically stable. Because the coatings are so thin, the only substantial changes to the original material are its new stability, hardness, and wear. Another company, Nanophase, manufactures powders of metal oxides with nanoscale dimensions. These materials can be used for surface coatings and for catalytic applications, and Nanophase can manufacture them by the pound or by the ton.

Perhaps the most important aspect of nanoscience and nanotechnology is the ability to respond to what might be called grand challenges. These are major problems such as diagnosing particular forms of cancer, stopping corrosion on metal bridges, providing early warning of heart malfunctions, developing environmentally friendly and significant new energy sources, providing total assurance of food safety, producing reliable long-term storage of information, and so on. It is in addressing these grand challenges that the greatest potential of the nanosciences to contribute to the betterment and safety of society can be uncovered.

More extensive discussions on the nature of nanoscience and nanotechnology can be found in *Nanotechnology: A Gentle Introduction to the Next Big Idea.* The overview in this chapter should be enough of a foundation to discuss security and how nanotechnology can help provide it. It is to this issue that we turn our attention in Chapter 3.

The New Battlespace

"Military applications of molecular manufacturing have even greater potential than nuclear weapons to radically change the balance of power." —Admiral David Jeremiah

The Changing Face of War

America possesses the strongest military in the world. In terms of manpower, equipment, training, and capabilities, its armed forces have distinguished America as perhaps the only remaining superpower. This military might has been crucial to America's prosperity—the possession of great wealth requires the capability to defend it. It has also been pivotal to the geopolitical stability of the world. After the invention of nuclear weapons in 1945 and the Cold War that followed it, America, Russia, and the other nuclear powers developed large, robust arsenals that could not be eliminated with a single preemptive strike. Any attack on one of these nations would, in principle, lead to nuclear retaliation and to the total devastation of all the parties involved. This military doctrine, sometimes called Mutually Assured Destruction (or, ironically, MAD), dominated strategic thinking for much of the Cold War. It also meant a shift in the methods and locations of war, since another world war is unthinkable to sane leaders in the context of MAD. Instead, as JFK pointed out, the theaters of war changed to Asia, the Middle East, Africa, and other less-developed areas. With the collapse of the Soviet Union and the end of the Cold War came a corresponding change in the threats to America. Most Americans (and citizens of other developed countries) are no longer concerned about invasion, as we were until World War II, or about extinction, as we were during the Cold War. In this era of globalization, we are concerned about terrorist strikes that result in economic or political instability as well as death. It's useful to look at why this shift occurred and at the change in thinking that has accompanied it in order to understand the new challenge to America's armed forces. While this discussion is necessarily simplified and highly abbreviated, it still adds context to the issues.

During the Cold War, America and the Soviet Union were reluctant to fight or even skirmish with each other directly since matters might escalate out of control. Instead they tended to arm and train local militias, insurgents, and armies in the nations they considered important to their national interests, such as Vietnam, Afghanistan, Cuba, Korea, Iran, and Iraq. Indigenous troops were taught how to use modern weapons systems and how to fight guerrilla wars. The superpowers also tended to back governments of convenience, generally defined as those that would be amenable to the superpower's own political or economic interests rather than those that were legitimately elected. This led to ongoing strife and a culture of violence that has caused many of these nations to remain hot spots to this day. While terrorism predates the Cold War, the overwhelming majority of today's most notorious terrorists, including Osama bin Laden, either come from hot spot countries or were trained there. These people (with the possible exception of colorful Iraqi information minister Muhammed Saeed al-Sahaf) are under no illusions that they are capable of fighting the United States directly to achieve their objectives. Instead, they are trained in unconventional warfare and, like good martial artists, apply force precisely at their enemy's weakest points, often using his own strength against him. This means killing civilians, toppling symbols and monuments, disrupting the economy, and where possible using the other side's own technology, as happened on 9/11.

In parallel to this, public sentiment in America has changed since 1945. During the world wars and all preceding conflicts, victory was paramount. Domestic casualties were minimized when practical, but a fairly high casualty rate was considered acceptable. Enemy casualties, both military and domestic, were hardly given notice. This was the psychology that led to the firebombing of Dresden and to the nuclear destruction of Hiroshima and Nagasaki. In the Korean War, and to a much greater extent in Vietnam, this began to change. Americans questioned why troops were being sent for dubious purposes to a faraway nation that most people had never heard of. The human losses became unacceptable. Enemy casualties have also now become a major issue when fighting in countries that sponsor terrorism. This is attributable to a few causes. Most such countries are not representative governments, and many feel that the citizens should not suffer for the sins of their leaders. Also, some feel the pangs of a guilty conscience, for the United States helped to create many of the situations that have led to the surge in world terrorism. While very few question that a country is justified in striking back against an attacker, the debate over the recent war in Iraq (especially Germany's position) has underscored the philosophical division on the policy of risking additional civilian deaths in an effort to stop terrorism.

While these changes are certainly profound, perhaps the greatest changing influence on the face of war is not directly geographic, political, or psychological, but economic. The spread of free markets as well as advanced technology such as transportation systems, computer networks, and telecommunications systems have resulted in an economy that grows more and more globalized. In 2003, Germany, France, and Russia all had financial interests in Iraq, which made their opposition to the war seem as much like economic pragmatism as true idealism. American policy was equally stigmatized by accusations that oil needs rather than security concerns were behind its actions. More generally, the economies of the developed world are linked so closely that a ripple in any one affects the others. Pain in South America has echoes in Russia and in Europe. A recession in America or Japan brings down the rest of the world. Free-trade regions that lower the barriers still more are cropping up in Europe, the Americas, Asia, and possibly even Africa.

America has the strongest economy as well as the strongest military on earth, so it must play its cards carefully. It is clear that war and instability are counterproductive to the global economy. Threats of conflict depress consumer spending, travel, international trade, investment, and stock prices, leading to bankruptcies, negative growth, and unemployment. These issues are of surpassing importance to America, which is a large part of why the first President Bush failed to win reelection even after prosecuting a successful campaign in the Gulf War. This has led military thinkers such as General John Sheehan to speculate that the core mission of the U.S. military is now not just to protect America, but to help maintain global stability for economic development. Many military leaders, including General Colin Powell, have become more dove-ish than their political counterparts, and it has already been largely forgotten that the cabinet post now called Secretary of Defense was not long ago called Secretary of War. Unlike in the days of empire, war no longer pays.

These elements form the challenge to America's military. If at all possible, conditions leading to war, such as the accumulation of weapons of mass destruction by rogue states or those that sponsor terrorism, must be detected and stopped before they become problems. When wars do happen, they will be fought in faraway places and harsh environments such as swamps, jungles, and deserts. Neighboring countries may not be friendly, so operations may need to be headquartered far from the front lines. American and civilian casualties are unacceptable. Enemies are skilled guerrilla fighters who are relatively well armed, don't always wear uniforms or move in units, and have made extensive use of unconventional weapons including chemical and biological ones. These enemies also have very few constraints: they are willing to use any weapon and any tactic necessary, and they seldom worry about casualties on either side. Finally, the wars must be short to minimize economic consequences, but peacekeepers may need to remain in the fields for years after a conflict.

This was, more or less, the scenario of the 2003 war in Iraq. Add to it the continuous press coverage that largely eliminates cheating or secrecy, and the Iraq mission seemed impossible even with DoD's half a trillion dollar budget. To its credit, the military stepped up to the plate and embraced the challenge. It made many changes to its organization such as relying more on intelligence and special forces groups, improving communications at every level, unifying command of domestic defense under the Department of Homeland Security, and increasing its capabilities for working in small, highly coordinated units. These changes are beyond the scope of this book, but they are an important first step.

The military was also very lucky. Despite concerns, no weapons of mass destruction were used—or even found—during the offensive in Iraq. If they had been, the situation might have turned out quite differently. Yet this is far from the only example of how vulnerable even the strongest military in the world can be in the face of its new challenges. The solutions to many of these problems are technological, not organizational. The core research operations of the Department of Defense, including the Defense Advanced Research Projects Agency (DARPA) and the Office of Naval Research, have issued lists of Grand Challenges. These are the crucial technological hurdles that must be overcome for the military to fulfill its new mission. Nanotechnology is key to the realization of almost all of these challenges, and the rest of this chapter is dedicated to describing many of them and explaining how they are being addressed.

Weapons of Mass Destruction: The Challenge

The use of chemical and biological agents as weapons of mass destruction is not a new phenomenon. People have been using biological weapons ever since they started throwing decomposing corpses and feces over city walls during sieges in hopes of causing the plague. Some of the first chemical weapons were poisons used on arrowheads. Chemical and biological weapons date back to prehistory.

Despite their early use, chemical and biological weapons never gained much popularity until modern times. Chemical and biological agents are generally difficult to "weaponize" or make into devices suitable for use in combat. Before missiles and bombs were common, this was a huge problem, since a plague represented a threat to attackers as well as to defenders when troops were closely mingled. In addition, lingering contamination would remain a problem for citizens and occupiers after the battle was over. Also, a change in wind or weather could blow a toxic cloud back at an attacker, and a storm or rain shower might reduce or negate its effects. Without a mechanism for delivering them to an opponent's homeland, chemical and biological weapons have virtually no value as weapons since they are so volatile and unpredictable. It is hard enough to protect an attacking army from such weapons. Until now, it has been virtually impossible to protect a civilian population, but nanotechnology may change this.

World War I, which saw the first large-scale use of aircraft and modern field artillery, was also the battlefield on which chemical and biological weapons truly reached maturity. Blister agents (also called mustard gas), nerve gas, and a litany of other horrors killed troops and civilians in unprecedented numbers. Today, toxins such as ricin and sarin have been developed to be deadly in quantities of much less than a gram. To put this in perspective, ricin is considered twice as toxic as cobra venom. These agents are ideal weapons for terrorists since none of the factors that make them poor military weapons apply to terrorist tactics. Terrorists do not consider the unpredictability of these weapons to be detrimental since their psychological effect is undiminished. They are not concerned about weaponizing them since they can easily use suicide

Agent	Lethal Dosage (LD 50 g/kg body weight)	Source
Botulinum toxin	0.001	Bacterium
Diphtheria toxin	0.10	Bacterium
Ricin	3.0	Castor bean
VX	15.0	Chemical agent
GB	100.0	Chemocal agent
Anthrax	0.004-0.02	Bacterium
Plutonium	1	Nuclear fuel

Figure 3.1 This chart shows the relative toxicity of some chemical weapons and biotoxins. The most powerful ones can kill in quantities of less than one-hundredth of an ounce.

bombers to attack people or to plant toxic compounds directly into a building's ventilation system or a city's water reservoir. Sarin was used in this way when the Aum Shinrikiyo Supreme Truth cult distributed it on the Tokyo subway in 1995, killing a dozen people and injuring thousands. Ricin also has a history—it was a ricin pellet in an umbrella tip that was used in the 1978 assassination of Georgi Markov, the Bulgarian writer and journalist.

Chemical weapons are easy to make from common materials. Ricin, for example, is derived from castor beans with the aid of some basic chemistry apparatus, all of which are available in any high school science laboratory. The British authorities have already caught suspects trying to make the substance in London and there is some evidence that it was recently tested in northern Iraq. Hydrogen cyanide (HCN) is another easily obtained chemical agent. It is mass produced for industrial uses such as tempering steel, dyeing, engraving, gold mining, and the production of some plastics. It is also the active ingredient in Zyklon B, the gas used in Nazi gas chambers. It is still used in parts of the United States for criminal executions.

Unlike explosives, almost all of which are based on materials with a chemical similarity (they contain nitrates), there is no single commonality among all chemical or biological weapons. Even nuclear weapons are fairly easy to detect via the radiation they emit. Radiological contamination can be accurately and quickly measured using a hand-held Geiger counter, but there is no similar device for chemical or biological weapons. This makes detecting them by any current technique extremely difficult, since most equipment relies on automating a single test. "Bomb sniffers" and the detection equipment used at airports will catch most common types of explosives by testing for nitrates, but most chemical and biological weapons will get right through.

Today, the best approach for determining if a chemical weapon is present involves a technique called spectroscopy. Spectroscopy uses a sample's absorption, scattering, and other interactions with particular frequencies of light to develop a chemical fingerprint. This is the same approach that astronomers use to determine the atmospheric composition of planets and stars. Unfortunately, it can require a reasonable amount of laboratory equipment and a good quantity of a sample (or a large-scale telescope in the case of stellar measurements). Results can sometimes take hours to obtain. Biological agents are no easier to detect. When anthrax-tainted letters were sent around the country in 2002, it took three full days to give the all-clear at most suspected sites. This kind of time lapse is unacceptable for troops in the field. If an area may be contaminated or a shell may contain chemical or biological agents, soldiers on the front lines will have been exposed before they can react, and troops coming up behind them have only minutes to prepare.

In Iraq, the army used pigeons and chickens as an early chemical weapon warning system in what some servicemen jokingly called Operation KFC (for Kuwaiti Fried Chicken). The idea is that these birds are more vulnerable to the effects of poison gas than people are. If they die, soldiers can conclude that there is a poison in the atmosphere. This was also the reason that miners took canaries into coal mines in the 19th Century. While better than nothing, the problems with this scheme are manifold. For a start, almost anything from blowing sand to heatstroke can kill a chicken, so there will be plenty of false alarms. More importantly, when much less than a gram of a toxin is enough to kill a person, by the time the chicken dies it is already too late.

Even assuming it were possible to detect chemical and biological toxins instantly and accurately, how could troops be protected from them? Currently, it is necessary to keep troops inside protected vehicles or to make them wear heavy, protective suits that cover the entire body and prevent anything from getting in. This is similar to trying to fight in a space suit, and it isn't practical for active combat or for any mission requiring more than a few hours exposure in desert or jungle conditions.

After the dust settles and the toxins have been detected, what options are there for destroying them and decontaminating an area? If the toxin is contained within a shell it is relatively easy to destroy, but once it is distributed, it is much more difficult. Some equipment can be power-scrubbed or -cleaned, but the process is expensive, time consuming, and results in a great deal of contaminated water. What if the contamination is in the desert where water is scarce? What if the water supply itself is contaminated? While some biological toxins can be neutralized by boiling or removed by micro-filtration, it is often hard to explain this process to local people in hostile territory, and many chemicals cannot be removed this way. Once they are infected, people can also transmit anthrax and other diseases used as weapons. How can people be tested quickly and quarantined before they infect others?

These scenarios are more likely than they seem. Saddam Hussein used chemical weapons on Kurds in his own country. He also used eco-terrorist tactics in the Gulf War by setting fire to oil wells. He might well have used chemical weapons to slash and burn his territory had he remained in power long enough after the invasion began. It is clear that if chemical or biological weapons had come into play in Iraq or do come into play before adequate countermeasures are developed, shock and awe military tactics will not help make the U.S. military any less vulnerable. Something much more defensive and more powerful will be needed.

Weapons of Mass Destruction: The Solutions

Part I: Detection

Enter nanotechnology. What if the desk-sized labs that could determine the presence of a chemical or biological toxin could be integrated onto a chip the size of a fingernail? What if they could also perform their tests in seconds or even fractions of a second for the more common toxins? This would make them lightweight enough for any soldier to carry, and quick enough to allow that soldier to take action before it is too late. This may be possible using a nanotechnology technique called lab-on-a-chip. Labs-on-chips are complete chemical or biological laboratories including chemical reactor cells, sample storage, and micro- or nanoscale channels, all integrated onto a microchip in much the same way that circuits are.

Companies such as Nanosphere and MicroSensor Systems have already developed the first few devices utilizing this approach for biotoxin detection. Right now these sensors are still the size of laptop computers, but they are getting smaller rapidly. There is reason to believe that thumbnail-size sensors will be possible since companies including Agilent (formerly part of Hewlett Packard) and Affymetrix have already made such small sensors for simpler kinds of analysis (Figure 3.3).

Figure 3.2 shows a sample device from Nanosphere, a company specializing in DNA-based sensor nanotechnology. How the device works is shown in Figure 3.4. The sensor contains nanoscale particles of gold (often called nanodots), and each of the gold nanodots has several short strands of specifically designed DNA attached to it. The DNA attached to the nanodots will chemically bind to a certain target DNA and to no other; for example, it might bind only to the DNA of anthrax. If anthrax is present, the DNA strands connected to the nanodots will bind to it, drawing the nanodots closer together. As

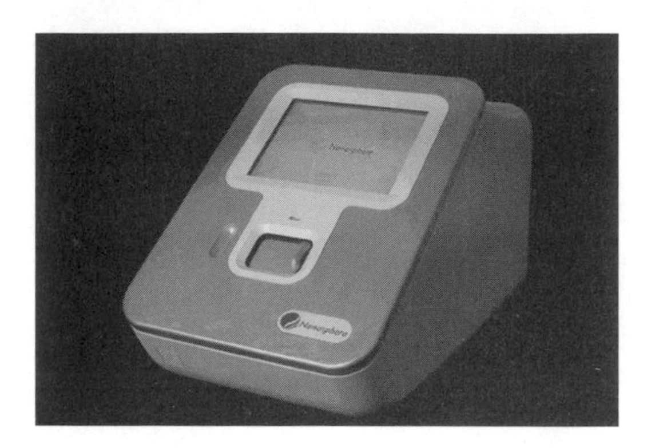

Figure 3.2 The Verigene ID device from Nanosphere is a sensor system for biotoxins. It can replace a small truckload of lab equipment. Courtesy of Nanosphere, Inc.

Figure 3.3 Picture of a GeneChip array. Courtesy of Affymetrix.

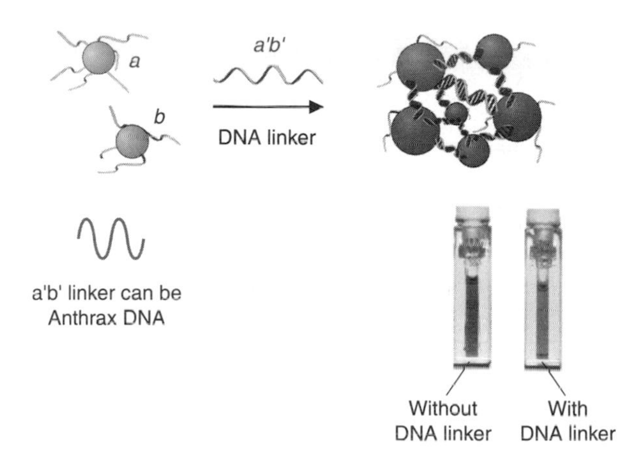

Figure 3.4 The upper schematic shows how the nanodots in a colorimetric sensor are brought together upon binding to the DNA target (in this case anthrax). The clustered dots have a different color than the unclustered ones in the photograph below. Courtesy of the Mirkin Group, Northwestern University.

they draw closer together, the nanodots undergo a change that only happens at the nanoscale—they change color without changing structure or composition.

The reason for the color change is complex and has to do with the coupling of the properties of quantum mechanics with the properties of the visible world, but the result is straightforward. In the presence of anthrax (or whatever biotoxin the sensor is designed to detect) and under no other conditions, the sensor will change color—in this case, from red to blue. The nanoscale color change may be detected by a laser and transmitted to a soldier via a computer or heads-up display. Alternatively, if a more direct reading is desired, the soldier could carry a visible quantity of these sensors, effectively a litmus paper for biotoxins. While the science behind it may be complicated, this is a test that anyone can use.

This technology and several other promising candidates are evolving quickly, and soon all that will be needed to develop a test for a new toxin is the toxin's genetic fingerprint. The same technology can also be used for diagnosing diseases. In the case of SARS, a genetic fingerprint was found within three months. As DNA sequencing technology continues to develop, this discovery period will get shorter and shorter. With a library of biotoxin fingerprints, a sensor for each toxin could be integrated into a lab-on-a-chip and a universal, nearly instantaneous, very sensitive, and essentially 100% accurate biotoxin sensor could be a reality.

Once a sensor the size of (or very much smaller than) a thumbnail is developed, it could be installed anywhere on a soldier's uniform. Some toxins are very volatile (they evaporate easily) and some are very stable. If a sensor is located on a helmet, for example, it would not be very helpful in detecting botulinus toxin in a puddle, whereas a sensor in a boot might not be as helpful in detecting airborne toxins. Ideally, dozens or even hundreds of sensors might be woven into a uniform and interconnected by tiny wires running throughout the fabric. This kind of matrix could still remain extremely lightweight and could also serve as a general communications mesh for all of the soldier's equipment, allowing many applications that we will discuss later. A primitive version of this kind of wire mesh is currently used in fencers' jackets to detect the impact of an opponent's sword during competitions.

An alternative to weaving sensors into a uniform is to move the sensors around so they are exposed to everything a soldier is exposed to. One plan for doing this involves injecting sensors into a soldier's bloodstream. The sensors would circulate through the bloodstream and could be monitored at a place where blood vessels are closest to the surface, such as in the eye. The monitoring unit could be installed in goggles or on a microphone boom of a communications headset. While quite invasive, these so-called *in vivo* sensors could also have other uses in continuously monitoring the health of a soldier.

Since soldiers on the battlefield are already linked by radio communications, any sensor solution should also be tied into the link, perhaps on a separate data channel. If each sensor on every soldier could intercommunicate, a whole unit could start reacting immediately to the presence of a threat as soon as a single sensor was triggered. Also, statistical analysis could help weed out any false positives, making a unit's awareness of threats as close to perfect as possible.

Part II: Protection

Once a toxin has been recognized, the next thing to do is to protect the soldier. Since some types of nanotechnology-based sensors can be accurate to the level of detecting individual molecules of a toxin, there should be some advance warning before enough toxin to cause fatality has been delivered. However, the warning could come only seconds before a lethal dose, far too short a time to put on a chemical defense suit—especially in the middle of a firefight. Fortunately, nanotechnology provides an answer for this, too.

Consider using nanotechnology-enabled smart materials for other purposes. Materials already exist that change their properties when exposed to different external stimuli. Human skin is a good example. It tans in the presence of direct sunlight, sweats when it gets too hot, develops calluses where it is frequently abraded, and heals itself when it is punctured. For these reasons, nanotechnology is of interest to Estée Lauder, Revlon, and others whose customers don't usually dress in combat fatigues. Artificial materials with skinlike properties have been created, but it is necessary to engineer at the molecular level to create what is required for chemical defense. The idea is to create a fabric that is a comfortable open-weave combat uniform under most conditions but, when exposed to a certain stimulus (an electrical signal from a sensor, for example), could restructure itself at the nanoscale to become an airtight shell. Air could be delivered via an oxygen rebreather of the type used by some divers or through nanotechnology-improved gas masks.

Such a uniform is not only possible, but likely. Nanotechnology-enhanced fabrics that are totally resistant to the penetration of liquids (and are thus stain resistant) are already made by companies such as Nano-Tex, whose fabric is used in khaki pants sold by Eddie Bauer and others (see Figure 3.5). While you might expect rubber or nylon pants to reject water and stains in this way (although why you would wear rubber pants is another question), only nanotechnology can combine the properties of comfortable cotton fibers with the desirable properties of repelling liquid to make fabric that is indistinguishable from ordinary khakis until you spill on it. Employing so-called smart material in soldiers' uniforms is just an evolution of the techniques already in use. As we'll see later, this is only one of the ways nanotechnology will affect soldiers' uniforms.

It is important to protect a soldier in his uniform, but better defenses are also required for buildings, vehicles, and other installations. Filtering out biotoxins is in some ways simpler than filtering out chemical weapons, since a single virus or bacterium is more than 100 times larger than a small molecule such as HCN or oxygen. This means that a filter with pores measuring tens of nanometers can be made to block out most biotoxins but to let air through. Unfortunately, these filters don't work well for chemical weapons where the difference in size between air and toxin may be very small indeed.

Figure 3.5 This photograph shows pants based on nanotechnology-enhanced fabric. The inset shows how the fabric resists the penetration of liquids. Courtesy of Eddie Bauer.

Right now, most chemical weapons protection is based on a similar technology to World War I–era gas masks. A simple air filter is impregnated with activated carbon (also used in commercially available water filters) and other chemicals that adsorb many common contaminants. This approach is far from effective in many cases. For example, HCN is one of many common chemical weapons that does not react much with activated carbon.

A different approach involves making an artificial membrane and permeating it with nanotubes. Nanotubes are shaped like drinking straws made of carbon, and their diameters can be controlled with subnanometer precision. Using this approach could create a molecular sieve that would block all but the very smallest molecules—oxygen and nitrogen, for example—and make a very good filter indeed. This effect depends on tailoring the sizes of individual molecules and is only possible through nanotechnology.

Part III: Remediation

If an area is contaminated with chemical or biological agents, it must be cleaned up in a process commonly called remediation. Remediation can be done in a variety of ways, including old-fashioned soaking with soap and water, forcing breakdown via chemical reaction, or exposing to extreme heat. Nanoscience offers a few ways to vastly improve the efficiency of this process.

Consider the remediation approach of breaking down a toxic substance through a chemical reaction. A material's surface area and its reactivity are directly related. The greater the surface area is, the greater is the exposure to other materials with which to bond and interact. This is one way in which catalysts (materials that increase the speed of chemical reactions) work. Grinding a substance finer and finer increases surface area and the nanoscale is the logical limit. A quarter ounce of nanoparticles may have a surface area as great as a football field.

Some powders, such as aluminum oxide and magnesium oxide, have proven effective in neutralizing chemical agents and are already in use by the army (see Figure 3.6). Reducing these particles to the nanoscale not only increases their effec-

Figure 3.6 This photograph shows soldiers using powdered sorbents for decontamination. Newer nanopowdered sorbents will make the process more effective and efficient. Courtesy of the U.S. Army.

tiveness by increasing reactivity, but also allows them to be impregnated into the fibers of filters like the gas mask filters we just discussed. Effectively this allows them to do what activated carbon did in older filters—aggressively bind to and remove chemical agents. While activated carbon does work this way, it is only one example of a whole family of nanoparticles that can be developed to deal with a host of chemical weapons. Activated carbon was created by trial and error rather than through a knowledge of the underlying mechanism that makes it function. However, research in nanoscience is revealing these mechanisms and is allowing the design of new nanoparticles specifically engineered to react with many chemicals that activated carbon can't handle.

destruction, we introduced the idea of making the uniform into a matrix to which sensors could be attached. This is only the beginning.

Protecting a soldier is important, but ideally he or she should also be difficult to see and therefore difficult to attack. This is the idea behind current camouflage uniforms, stealth bombers, and the ability to fight effectively at night—advanced night vision gear is already based on nanotechnology and represents an early application.

The camouflage in current uniforms is called passive camouflage, meaning that it doesn't change or adapt to different circumstances. Active camouflage, however, is just the opposite—it blends in with whatever environment it is exposed to, just as a chameleon blends in with rocks, leaves, or soil. A chameleonlike approach to making soldiers very difficult to detect may be possible through nanotechnology and biomimicry. By working at the same scale as nature, natural processes like those of the chameleon can be emulated and integrated into useful devices. James Bond fans recently saw an exaggerated version of active camouflage in *Die Another Day*, in which 007's car could disappear into the ice with the flick of a switch. While active camouflage probably won't be good enough to fool someone up close, active camouflage could certainly have applications for jets and aircraft, making them as hard to see with the eye as they are hard to see with radar.

While it is best to avoid injury through stealth wherever possible, it is necessary to minimize the damage a soldier sustains in an attack should one occur. One approach to doing this involves improving Kevlar, the fabric used to make most bulletproof vests. Kevlar is a polymer whose shape is rigid as opposed to the more pliable, thread-like structure of most polymer chains. This rigidity allows Kevlar to disperse energy from local impacts pretty well and to prevent bullets from getting through. Very new smart materials being developed by Ray Baughman's group at the University of Texas in Dallas incoporate nanotube fibers into an open-weave cloth that is more than four times tougher than spider silk and seventeen times tougher than Kevlar.

Timothy M. Swager
Professor of Chemistry and Associate Director
Institute for Soldier Nanotechnologies

To protect humans from ballistic impacts is difficult. What is needed is a material with the flexibility of cloth, but that instantaneously can become stronger than steel. The design standards for high strength and ballistic protection materials such as Kevlar were developed decades ago. They involve polymers that behave like rigid rods, effectively glued together by weak molecular-bonds. Nanoscale assemblies offer the best prospects for generating materials with the needed revolutionary properties. Translation of macroscopic mechanics to nanometer dimensions such as the formation of molecular analogs of chainmail—a concept from the middle ages—can be used to create needed flexibility while maintaining a material's strength.

Being able to smell danger would be a powerful tool against many threats. Realizing this potential requires sensors with vastly enhanced sensitivity. Molecular and nanoscopic wires can transport electrical charge and excited states over many molecularly responsive switching sites. The power of nanoscopic circuitry is illustrated in sensors that rival the detection limits of trained canines and allow for the detection of buried landmines based upon the smell of explosives. These sensors could be 100,000 times more sensitive than the state of the art detectors used in airports today.

Researchers at MIT are pursuing another approach. This involves exploring the potential of making a uniform that not only reacts to chemical or biological toxins, but can stiffen and act as armor against ballistic threats such as bullets and fragmentation. Imagine a sheet of paper held at one end. It is floppy and flexible, but if you curve it slightly as you hold it, it becomes stiff and rigid. One approach to allowing a uniform to act as armor involves packing nanoparticles of iron onto the fabric's fibers. Normally these particles do little other than add a bit of weight, but in the presence of a magnetic field they align and force the fibers to become rigid. The rigid fibers could offer significant protection against ballistic threats and the armor could be activated or deactivated for convenience, depending on whether the soldier is in battle or just patrolling. While iron nanoparticles may add a bit of weight, they will be significantly lighter than the 15-pound vests that are common today.

Unfortunately, while improved armor will reduce injuries, even nano-protected soldiers can still be hurt in battle. In recent conflicts, some 50% of battlefield fatalities were not immediate but were the result of bleeding or complications occurring several minutes, hours, or even days after a wound was inflicted. Nanotechnology may help to reduce this problem.

If a soldier is hurt in the middle of a firefight, he or she must wait for some kind of assistance from a corpsman or a comrade with a medical kit. What if the nano-enhanced uniform could help with this, too? Sensors could be woven into the fabric that could look for hemoglobin, the oxygen-carrying compound in the blood. Good hemoglobin sensors have already been developed for medical diagnostics such as the detection of blood in stool samples, and these could be incorporated into labs-on-chips and used to begin treatment. Along with the hemoglobin sensors, the tiny labs could also contain supplies of antiseptics, antibiotics, and anesthetics. The fabric itself could contain a smart material liner that would adhere to the wound and act as a temporary bandage.

Groups like the Institute for New Materials in Germany have already demonstrated using biocidal nanoparticles (particles that kill bacteria on contact) to keep medical equipment such as hearing aids sterile. These nanoparticles aren't made of anything exotic—silver is one of the best nano-biocides. It has been suggested that these kinds of biocidal particles could be incorporated into hospital bed sheets to reduce the incidence of cross-infection, and they could be used in uniforms too. Wearing a biocidal uniform could help reduce infection from injuries as well as reduce other kinds of contamination that can occur when uniforms cannot be changed or cleaned regularly. It is not uncommon for a soldier in the battlefield to go for several days working hard, sweating and perhaps bleeding, without having an opportunity to change clothes or visit the laundry.

We began this section with a criticism of a soldier's kit. We described it as too cumbersome, and too bulky. We accused it of a lack of integration. Yet now we are suggesting adding a wire mesh, hundreds of sensors, iron nanoparticles for protection, silver nanoparticles for the reduction of infection, nanoscale medical kits, and who knows what to enable active camouflage. It sounds like we've made the situation rather worse than better, but this is not the case.

First of all, these technologies are not as unrelated as they seem. They should all be integrated into a single, coherent system—the next generation uniform. This system will contain a central communications channel in the form of a wire mesh or similar solution and integrated processors to handle information. It will contain functional components or subsystems that react to specific threats. Many of these subsystems have multiple applications. A sensor can be used to detect an external threat, but it may also be used to detect blood and apply treatment. Clothing that can change from flexible to stiff may function as armor or as a cast or tourniquet for medical purposes. One approach to integrating all of these solutions into a uniform involves placing nanoscale channels into the core of the fibers within the fabric (see Figure 3.7). Not only would this solution enable the applications we've discussed, but the electromechanical channel opens the door to using the uniform to actually enhance human performance. It might, for example, augment human muscles and make the wearer effectively stronger, thus providing a unique solution to the weight problem.

Whether they enhance performance or not, weight may not be an issue for such a uniform. To understand why, remember just how small the nanoscale is. Ask anyone who has worn a pair of Nano Care pants if there is any difference in weight between them and ordinary khakis. Ask anyone who has used a Wilson Double-Core tennis ball (based on nanotechnology-

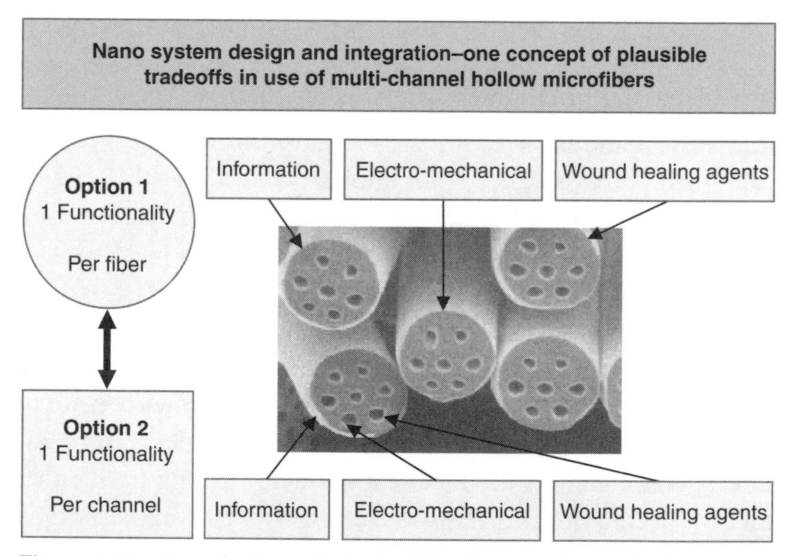

Figure 3.7 A multichanneled fiber like the one shown in the picture might be used to integrate the systems in a soldier's uniform. Courtesy of DuPont.

enhanced composites) if there is any difference in weight between them and ordinary tennis balls. See if you can feel the difference in weight between a microchip with copper technology and one based on aluminum. While it isn't certain, it is likely that one or more of the improvements we have talked about will be possible without affecting the weight or comfort of a uniform except in a positive direction.

Ground forces tend to remain in trouble spots long after the Air Force, Navy, and Marines have pulled out. They act as police, peacekeepers, and even engineers in the restoration of basic services after fighting is over. They need a better layer of lightweight protection to help them get these jobs done. Nanotechnology provides the only likely solutions to these problems.

The Man-Machine Interface

The tasks of modern soldiers might well be called superhuman and thus require superhuman characteristics to accomplish them. We briefly mentioned using channels in the fibers of uniforms to provide performance enhancement for soldiers, but a lot of other approaches are possible as well.

One performance enhancement possibility for increasing strength and reflexes, as well as flying aircraft or driving cars, is a direct interface between the human nervous system and electronics. There have been many promising results in this direction, particularly in the area of prosthetics and bionics. Hearing aids, for example, used to work by simply amplifying the sounds around them. Neuro-electronic hearing aids work by actually transforming (technically the term is transducing) the sound into a neural impulse and injecting it into the nervous system, bypassing the eardrum entirely. Similarly, neuro-electronic eyes (Figures 3.8, 3.9, and 3.10) attempt to project images directly onto the retina. These devices were created to solve the most

Figure 3.8 A two-millimeter ASR (Artificial Silicon Retina) chip lying on a penny. This chip can be embedded in the eye and may help to restore certain kinds of vision loss. Courtesy of Optobionics, Corp.

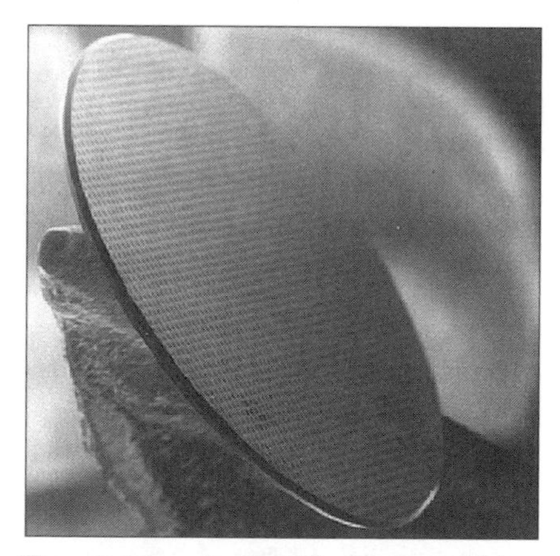

Figure 3.9 A magnified image of the ASR microchip shown in Figure 3.8. Courtesy of Optobionics, Corp.

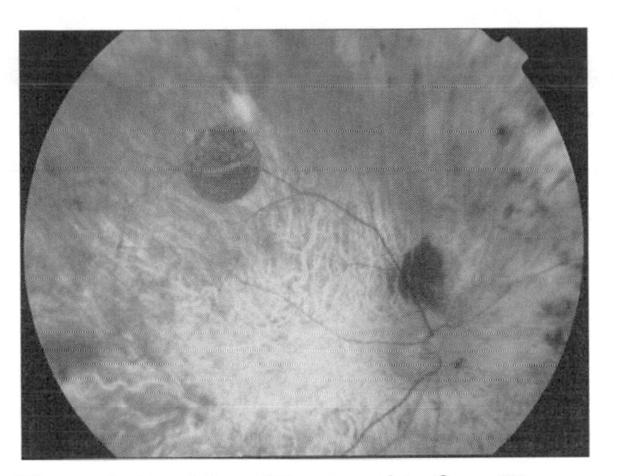

Figure 3.10 The ASR microchip from Figure 3.8 implanted in the human eye. Courtesy of Optobionics, Corp.

untreatable forms of blindness and deafness and are having good results. Optobionics, a company that develops vision solutions, has demonstrated that some trial patients who could see only darkness now see blurry shapes and those who saw blurry shapes can now distinguish between teams at a ball game.

This kind of interface promises treatments for all sorts of conditions from missing limbs to quadriplegia. As it becomes less invasive it could be used to eliminate the old-fashioned steering yoke so pilots can fly planes directly (giving the expression "fly-by-wire" a whole new meaning) or to control servo-motors in a battle suit which could increase speed, strength, protection, and more. Working at the scale of nature will make this possible.

Out of Harm's Way

Perhaps one of the most controversial ideas for protecting soldiers involves taking them out of combat. This does not mean that wars and conflicts can be entirely avoided, but it may be possible to use more and more mechanization to fight future wars.

Consider the Predator UAV (Unmanned Aerial Vehicle) (Figure 3.11). The Predator is just one in a family of unmanned aircraft produced by General Atomics, but in 2001 it made history as the first unmanned aircraft to launch a live antitank missile in combat. While not strictly based on nanotechnology, the Predator shows an interesting trend. Remotely controlled combat aircraft are now a reality, as are autonomous missile systems like the Tomahawk cruise missile, which can find a target using GPS and other programming even over complex terrain. Indeed, missiles guided by lasers and heat are now old hat.

Ground vehicles, however, are not yet automated. Unlike Predator or Tomahawk, the missions of unmanned ground vehicles do not involve simply going to a location, attempting a kill, and, in the case of Predator, returning to base. A tank or other ground combat vehicle must negotiate changing terrain and deal with rapidly changing threats. Even the short period of latency (the time between when a signal is sent and received) between it and a remote controller could make a world of difference. This is very similar to the set of challenges

Figure 3.11 Photograph of the UAV (Unmanned Aerial Vehicle) Predator. Courtesy of General Atomics Aeronautical Systems, Inc.

NASA faces in creating landing vehicles for Mars. To be successful, it will have to be somewhat autonomous, and it will likely rely on several forms of information technology and processing that are not yet fully developed. These include image recognition and, more controversially, artificial intelligence.

It may be a long time before the military or the public is ready for unmanned vehicles or even robots accompanying soldiers into battle, but DARPA (the Defense Advanced Research Projects Agency) and others have been looking at the problem for some years. See Figure 3.12.

Several of the components required to make autonomous land vehicles will come from nanotechnology. These include high-performance computers using new architectures that may make artificial intelligence a reality. This artificial intelligence will not necessarily resemble human intelligence, but it will more likely be a specialized system capable of making intelligent decisions of a certain type, a sort of expert system. Thus far even the most advanced silicon computers have been able to show only modest results in this regard since they are designed on an architecture optimized for numerical computation, not cognitive intelligence. Nanotechnological approaches to computing

Figure 3.12 A concept unmanned ground combat vehicle. Courtesy of System Planning Corp.

including the use of molecular electronics, bioelectronics (electronic components integrating proteins and other biological entities), and quantum computing may change this.

Better, Faster, Tougher, Smarter

While autonomous robotic land vehicles may still be in the future, nanotechnology has changes in store for conventional combat vehicles, such as fighters, helicopters, and tanks. Most of these changes will come from the application of nanotechnology-based smart materials.

For example, consider stealth aircraft. While the most advanced models are said to have radar profiles comparable to that of a bumblebee, they make large sacrifices in terms of design. The first stealth fighter, the F-117, shown in Figure 3.13, has an exotic, panel-shaped design and is far from being the most maneuverable or speedy fighter in the sky. The F/A-22

Figure 3.13 Photograph of the F-117, one of the first stealth fighter aircraft. Courtesy of Lockheed Martin Aeronautics Co.

and other later stealth designs have improved on the F-117, and nanotechnology will help still more. Nanoparticles with massive absorption cross sections for radar and infrared are being developed. This means that they soak up a huge amount of energy and can therefore prevent detection. In addition, the same active camouflage we discussed for soldiers' uniforms may be applied to aircraft and ground vehicles.

Nanotechnology will also make vehicles faster. Argonide Coporation already produces Alex, a rocket-fuel additive based on nanoparticles of aluminum. Alex vastly increases the efficiency of burning hydrocarbon fuels like kerosene and could have applications for jet fuels. Such an additive could also increase the efficiency of powder burn in a rifle cartridge. Figure 3.14 shows the improvement that Alex makes possible.

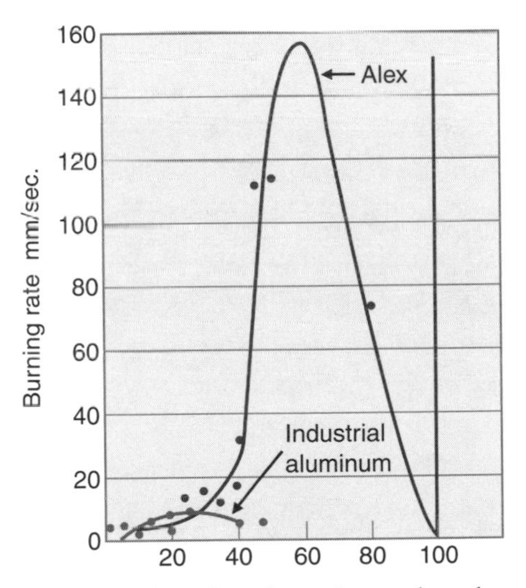

Figure 3.14 This chart shows the relative performance of Alex, a nanopowdered fuel additive, and its microparticle competitors. Data courtesy of Argonide Corp.

Another barrier to speed, fuel efficiency, and deployment is the weight of combat vehicles. A tank may weigh 60 tons and infantry vehicles weigh almost as much. Lighter materials such as the ever-useful carbon nanotube may help solve this problem. Next-generation commercial aircraft already utilize carbon fiber and other superstrong, lightweight carbon composites in their construction, but none of these materials can come close to the prowess of nanotubes. With strength on the order of 60 times that of steel at a fraction of the weight, nanotubes will be used to make aircraft that are faster and that require shorter runways and less fuel. Nanotubes or other such nanomaterials can even be used to armor aircraft such as attack helicopters which are usually lightly armored for weight reasons. Companies like CNI and NanoLab are now offering nanotubes in appreciable quantities, but they still can cost about a hundred dollars a gram. Still, this price is far lower than it was just two years ago, and two years before that the idea of bulk nanotubes was still a pipe dream.

While carbon nanotubes may not yet be the armor of choice, nano-enhanced armor is already in use protecting troops. The SAG (Save A Gunner) turret, shown in Figure 3.15, for light combat vehicles such as HMMWVs is a case in point. Produced by US Global Nanospace, it uses nano-fiber materials to create a gun turret that is less than 200 pounds, has minimal impact on mobility, and yet can withstand six successive 7.62-caliber bullet strikes in the same spot (according to the company). US Global Nanospace uses similar technology to make blast-proof doors for aircraft and materials for dealing with explosives.

NASA engineers have started thinking about what its next-generation orbiter might look like, and it also incorporates many of these ideas. In addition, NASA scientists propose using nanotechnology-based coatings to make the skin of a new spacecraft almost friction free, reducing heat from reentry. They are also considering the idea of self-healing materi-

Figure 3.15 Image of a SAG (Save A Gunner) turret made out of G-LAM nanofiber. Courtesy of US Global Nanospace, Inc., and Rich Schineller.

als—these are materials that can start to repair themselves if they are damaged in the same way that human skin does. Self-healing materials often operate on a similar premise to skin. Some of the first self-healing plastics have been developed at the University of Illinois. They work by incorporating microspheres of liquid monomers (the building blocks of plastic polymers) and a catalyst into the plastic matrix. These microspheres correspond to platelets in human blood. When a crack forms in the plastic, the microspheres break to expose the monomers and the catalyst, which then move in to seal the crack just as platelets move in to seal wounds in skin. Self-healing materials on aircraft could possibly have prevented the crash of American Airlines flight 587 in New York, the crash of Air France Concorde flight 4590 in Paris, and even the explosion of the space shuttle Columbia. Aside from their applications for civil aviation, they could also make combat aircraft much more survivable.

Batteries Not Included

Alternative energy sources will be necessary to take advantage of many of these new nanotechnologies. Current equipment is already power-hungry, and many of the technologies we've discussed will exacerbate the problem. Soldiers currently carry more than 20 pounds of batteries and portable power supplies into battle, but consider the issues of other kinds of combat equipment. Ground vehicles and aircraft are powered using hydrocarbon fuels (usually diesel or kerosene, although some multi-fuel engines can also run on gasoline or even alcohol). These fuel supplies are very volatile and can result in "secondary explosions," blasts that occur after an attack usually as a result of hazardous cargo such as fuel or ammunition. Fuel depots represent huge targets for the enemy and transporting fuel is a logistical nightmare.

Naval vessels have dealt with this problem in many cases by using nuclear fission reactors. These have the advantage that they seldom need to be refueled (only every few years) and thus give the vessels effectively unlimited operational range, but they also create other hazards. If a nuclear vessel is disabled, the fissionable material may be recovered and used for making a bomb. If it is destroyed or damaged the radioactive materials it contains could cause an environmental catastrophe.

Promising nanotechnology-based energy solutions such as fuel cells, efficient solar power, and even nuclear fusion (which has a nano component) could reduce or eliminate these concerns, and we'll look at these in Chapter 5.

The Tip of the Iceberg

The applications we have discussed in this chapter represent only the beginning of the possible applications of nanotechnology for defense. Nanotechnology will greatly increase the survivability of soldiers and equipment, especially for peace-

Debra Rolison
Naval Research Laboratory
Washington, DC

The importance of nanoscale architecture on multifunctional materials for energy storage and power generation

The equation of state that governs the ability of the Intelligence Community or the Departments of Defense or Homeland Security to counter terrorism, succinctly states:

no power = no mission

And while power requirements are mission-specific and may range from microwatts to megawatts, many missions require portable power as supplied by electrochemical power sources such as batteries, fuel cells, supercapacitors, ultracapacitors, or photovoltaics. The critical component in all of these devices is the electrified interface that mediates the ion, molecular, and electrochemistry necessary to create the current that runs sensors, robotics, laptops, and communication devices. Increasing the surface area that can be electrified, while retaining free access of ions and molecules to the electrode, yields higher capacitance in supercapacitors, fuller utilization of fuel in fuel cells, higher discharge rates in batteries.

Nanoarchitectures constructed by combining electrically conductive nanoscopic solids with nanoscale pores and channels amplify incredibly the nature of the surface, while ensuring molecular transport paths. The exploration of such multifunctional composites of "being and nothingness" is increasing our ability to rethink the design of materials and structures for a new generation of electrochemical power sources with higher performance.

keepers in the period after the hottest fighting is over and when a full combat rig is impractical. It will provide stealthier, tougher, and faster combat vehicles and increase battlespace information and awareness. It will improve communication within units and allow a soldier's kit to become an efficient, integrated system. It will combat the threats of chemical and biological weapons. It will be a crucial tool for weapons inspectors and those looking for weapons of mass destruction. It will provide lighter, more efficient portable power. It already helps our troops to see at night and provides some of the few chemical and biological defenses that we have so far developed. For these reasons, nanotechnology is key to the military's new mission.

When he developed one of the first machine guns during the Civil War, Richard Gatling sincerely hoped that the destruction it would wreak would be so great that no one would dare to use it and wars would become impossible. Clearly this did not happen, and the advice that guided one of his successors, Hiram Maxim, seemed much more apt: "If you wanted to make a lot of money, invent something that will enable these Europeans to cut each other's throats with greater facility." Perhaps Gatling's vision was most nearly attained by the nuclear weapon, but even the MAD doctrine depends on sane and stable governments having exclusive control of weapons of mass destruction. There are no credible claims that nanotechnology will take the final step in making war obsolete, but one thing that makes nanotechnology particularly exceptional among technologies with military applications is that it is primarily useful for defense. It can enhance soldier safety and capabilities, limit the destruction necessary to accomplish objectives, provide information and intelligence, coordinate troops, enhance reaction times, and even help heal the wounded. It may also make a lot of money. While it certainly could be used for weapons (we'll talk about this more in Chapter 6), nuclear weapons are still the ultimate killing machine. Since the American military already has destructive weapons of almost limitless capability, what it needs now is better defense and better capability for using its offensive weapons effectively. That is where nanotechnology will start.

Homeland Security

"Terrorism is the nuclear bomb for poor people." —Pablo Escobar

The State of the Nation

In the past, cities were walled and fortified to withstand attack from the outside. Settlement sites were selected not just for industrial purposes or for access to transportation as they are now, but because they were defensible. In today's world of bombers, missiles, tanks, and other mechanized combat equipment, the idea of fortifying a city seems to be obsolete. But this is not the case. Though walls may no longer do the job, modern defenses and security measures exist to harden critical infrastructure and lessen vulnerability to terrorist attacks.

Buildings are often engineered to withstand earthquakes and other predictable disasters (the World Trade Center towers were actually designed to withstand the impact of a small aircraft), but the issues of bomb resistance, air filtration, and threat detection are only modest concerns in the planning of most cities and structures. Excepting the construction of fallout shelters, hardening against the threat of nuclear strikes has been considered totally outlandish and not worth consideration. This perception must change. In a world tainted by terrorism, the buildings likely to be targeted can and should participate in their own defense.

Although major office buildings have been the targets of the most deadly terrorist attacks in America, more critical infrastructure often goes virtually unprotected. The minimal monitoring of water stations, mail sorting offices, power distribution points, and even vital telecommunications facilities make them highly vulnerable to chemical or biological attack. Despite the government's emphasis on airport security, airports are not much better, and bridges and train stations have essentially no security of any kind. Finally, the people required to respond first if any of these places are threatened—firefighters, police, and paramedics—need better protection on the job. Lives depend on their having it.

As always, the best way to defend against attack is to prevent the attack from occurring. The FBI and other intelligence agencies have been doing an impressive job since 9/11 of predicting attacks and finding culprits, but they need tools that allow them to extract information without unreasonable intrusion into innocent people's lives. These tools include better technology for detecting qualified threats and more effective decryption for intercepting communications. These technologies will require oversight, which we will look at in Chapter 6, but that is no reason to delay their development.

Terrorist attacks are the kind of disaster that America now dreads most keenly. Unfortunately, it seems likely that another major attack will occur on American soil. Nanotechnology will help mitigate the damage caused by such an attack, though only changes in policy that are beyond the scope of this book can eliminate the threat entirely. Other types of crises unrelated to terrorism could also be addressed by nanotechnology. Fires, floods, natural disasters, and the outbreak of disease are all areas where nanoapplications can make a significant positive difference. So nanotechnology for homeland security may have even more dual-use applications than nanotechnology for military applications, and it is easy to see how it could directly impact all of our lives.

Hardening the Hearts of Cities

In recent years, the most harmful terrorist attacks on America have been explosive attacks on large buildings. The first World Trade Center bombing, the Oklahoma City Federal Building bombing, and the 9/11 attacks all involved the destruction of office complexes. Commercial landlords and civic authorities reacted to the 9/11 attacks by implementing cursory inspections for vehicles pulling into parking lots under high rise buildings and by imposing more vigorous campaigns to check visitors' photo identification. While these precautions may help, how can parking attendants hope to find bombs that may be concealed in a suitcase or even under the seats of an SUV? How many would recognize a bomb even if they found it? And what prevents a terrorist from forging a photo ID?

Some of the sensor technology we discussed in Chapter 3 could be applied to inspecting vehicles, but a bomb can also be detonated from the street in front of a building and a car containing a bomb may run into or through a barricade. Individually harmless chemicals can be brought into buildings on separate trips, then mixed therein to produce disastrous weapons. How can office and apartment buildings be hardened to reduce the impact of such attacks? Nanotechnology offers several solutions that could be integrated into the design of future, attack-resistant buildings.

Start by considering structural issues. In San Francisco and other key earthquake zones, buildings are designed to be strong but also flexible. Some designs utilize a flexible central mast to support the rest of the building. Constructing these masts of nanocomposite materials or using nanocomposites to replace today's ubiquitous structural steel and reinforced concrete could significantly improve buildings' performance and protection.

The structure of a nanocomposite material resembles that of a normal composite material such as reinforced concrete, but shrunk to the nanoscale. Reinforced concrete consists of ordinary concrete poured over a steel mesh called rebar. It is formed at the macroscale, but it still manages to combine the hardness and compressive strength of concrete with the yield strength of the rebar to create a material with properties superior to either of its components alone. Nanocomposites could work the same way, but at the molecular level. For example, a nanocomposite might be made of a high-strength plastic wrapped around a rebar of nanotubes. Such a material could vastly outperform conventional construction materials, and a building based on it could withstand a much stronger explosion than anything that exists today.

Explosions typically cause damage in two ways. The first is through the blast concussion. The second is through heat, which is what actually caused the collapse of the World Trade Center towers. (The impact of airplanes alone would not have toppled the towers, but the immense heat generated by the burning jet fuel weakened key girders and supports.) Advanced nanomaterials can be used to deal with this problem. Figure 4.1 shows an example of a new kind of fire protection glass that is now becoming available. Buildings incorporating this and other nanomaterials now under development will have a much greater chance of surviving the heat of explosions and other kinds of fires.

Because of their immense strength, the amount of these nanomateirals required to build a nano-enhanced building is actually much smaller than the amount of conventional materials that would otherwise be used. This opens up whole new

Figure 4.1 Photograph of a new kind of fire protection glass in action. Courtesy of the Institute for New Materials.

design possibilities, and the nano school of architecture may not be far away.

Smelling Smoke

Buildings and homes are already equipped with sensors for many common hazards. Almost all cities have codes requiring smoke alarms and carbon monoxide detectors. Even burglar alarms are a kind of sensor, though for a very different sort of problem. Continuous environmental monitoring for chemical and biological weapons may never make sense for individual homes (though concerned citizens will certainly have the option to install such systems), but it makes a great deal of sense to put sensors in the air and water handling systems of large office and apartment buildings.

Most modern high-rise buildings already contain complex air handling systems for central air conditioning, heating, and humidity control. In many of these buildings windows cannot be opened for climate-control efficiency and safety reasons, but that makes them more vulnerable to attack. If a toxin were introduced into the air supply, getting clean air would require breaking the windows (causing another hazard in the form of falling glass), and clearing the building afterward would be extremely difficult.

If nanosensors were installed in high-rise air handlers and if the same kinds of filters we looked at for next-generation gas masks were introduced into the systems, the possibility for contamination would be reduced significantly. A building could be evacuated if a toxin were detected and air could be safely filtered in most cases. During day-to-day operations, these detectors and filters would be low cost and low maintenance. They might also have the beneficial side effect of extracting ozone, particulates, and other common pollutants from the building's air. In this case, preparing for disaster would also mean improving air quality all of the time.

These same kinds of hardening processes could also be implemented for key infrastructure like water, electricity, subways, mail centers, and communications. In these cases, the upgrades are not really optional. To fail to equip a water pumping and filtration station with state of the art sensor technology, for example, would be simply negligent. Remediation in the event of disaster would also have to begin at these points, so it makes sense to keep a supply of common remediation compounds on hand.

First Response

The term *first responders* has grown to mean police, firefighters, paramedics, and all other emergency workers who represent the front line of defense in case of a terrorist attack within the United States. These men and women face above-average risks in their day-to-day work, and the additional threat posed by terrorists both exacerbates certain basic problems and introduces new ones.

Like soldiers, police officers are at risk from enemy fire and often have to wear cumbersome protection including bulletproof vests while performing their duties. Firefighters require full sets of fire retardant suits, portable oxygen supplies, and axes, but despite these precautions even the best-protected firefighter can only remain in a burning building for a few minutes. Paramedics routinely deal with sick and contagious people, which is essentially a lower level biological attack. What can nanotechnology do to help?

Clearly the advanced-armor nanotechnology for soldiers (see Chapter 3) and heat-resistant nanotechnology for buildings will help first responders. Chemical-defense nanotechnologies will be useful not only for their stated purpose, but also for use in filtering smoke and various toxic chemicals from building fires. Nanotechnology sensors may also be useful in replacing canine units for many purposes—even a dog's impressive sense of smell cannot detect a single molecule of something. Sensors will certainly have applications for diagnosing people and determining the threat level of a situation.

Nanotechnology also has some specialty applications for first responders. When such professionals discover a bomb, the protocol is to remove and safely detonate it if possible. This may require transporting a device and taking some risk of detonating it by mistake. To help reduce the impact of such an accidental explosion, nanomaterials like US Global Nanospace's Blast-X can be integrated into wall panels to significantly reduce the effect of an explosion. Such materials might also be used as blast retardants in trucks, carrying cases, and even specially modified body armor.

First responders also need remediation agents (chemicals that can break down nasty molecules or biotoxins) at least as much as the military does. One interesting solution to the problem involves using photocatalytic self-cleaning nanolayers (sometimes called PSC layers). These are surface coatings consisting of a nano-thin layer of material (sometimes something as simple as silver titania) that can break down various harmful contaminants when exposed to light. The National Technology Transfer Center's Emergency Response Technology Program claims that anthrax, smallpox, botulinum toxins, ebola, cholera, bubonic plague, nerve agents, mustard agents, hydrogen cyanide, tear gases, and even exhaust fumes and smoke can be broken down by exposure to PSC layers. These nanolayers could find applications in chemical-defense masks, air handling systems, and self-cleaning lighting, paint, glass, and other household fixtures. PSC technology has been developed by PPG, GE Lighting, Pilkington Glass and a few groups in Japan, where the idea has been championed by Akira Fujishima of the University of Tokyo. An example of how it works is in Figures 4.2 and 4.3.

There are other promising technologies for first responders. Many of them are very similar to the soldier nanotechnologies we have already discussed (although missing the active camouflage and weapons kit).

Clean It Up!

Decontamination following either an attack or an accident is a significant challenge in homeland defense. If prevention fails, first responders can deal with the immediate issues of safety and containment, but the threats of dispersed toxic chemical or biological agents require broad remediation and decontamination strategies.

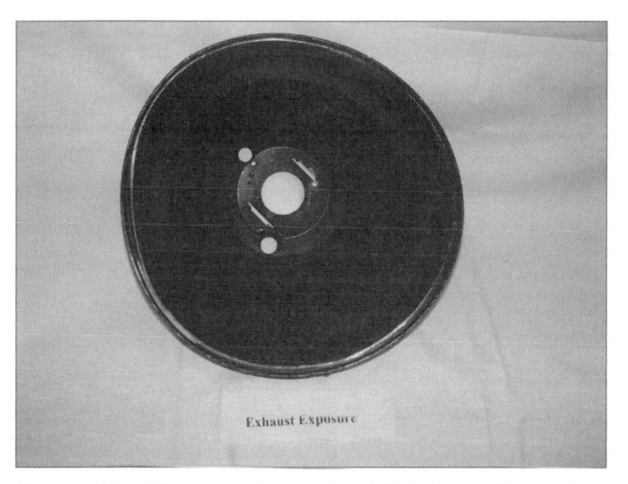

Figure 4.2 Photograph showing a lighting reflector that has been exposed to exhaust and other staining. Courtesy of Fred Simmons, SHC, Inc.

Figure 4.3 This photograph shows the same lighting reflector in Figure 4.2 after the photocatalytic nanolayers on its surface have been exposed to light and it has cleaned itself. Courtesy of Fred Simmons, SHC, Inc.

We've already noted that chemical reactions occur at surfaces, and that nanoparticles have huge ratios of surface to volume because of their tiny size (half an ounce of nanopowdered alumina has more surface area than a football field). So nanocrystals and powders will react with toxic agents far more rapidly and effectively than traditional crystals or powders. This suggests that decontamination with such powders can be rapid and complete. The powders can either adsorb the toxic species (much like World War I gas masks that used activated carbon) or react with and chemically destroy the toxin (like the PSC layers discussed above). The powders can actually be sprayed on or applied with special mittens being developed by the Army Research Laboratory.

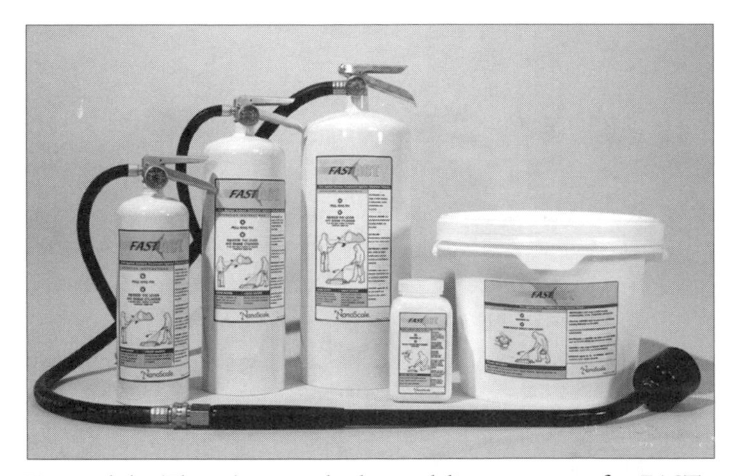

Figure 4.4 This photograph shows delivery systems for FAST-ACT (First Applied Sorbent Treatment Against Chemical Threats), a nanotechnology-based chemical hazard containment and neutralization system which is now available.According to NanoScale Materials, Inc., its manufacturer, FAST-ACT is the safest, most effective dry powder decontamination technology available today. Courtesy of NanoScale Materials, Inc.

Figure 2.1
Image of a single-walled carbon nanotube, viewed from the inside. Each of the bonds is roughly 0.14 nanometers long. *Courtesy of Chris Ewels, www.ewels.info.*

Agent	Lethal Dosage (LD 50 g/kg body weight)	Source
Botulinum toxin	0.001	Bacterium
Diphtheria toxin	0.10	Bacterium
Ricin	3.0	Castor bean
VX	15.0	Chemical agent
GB	100.0	Chemocal agent
Anthrax	0.004-0.02	Bacterium
Plutonium	1	Nuclear fuel

Figure 3.1
This chart shows the relative toxicity of some chemical weapons and biotoxins. The most powerful ones can kill in quantities of less than one-thousandth of an ounce.

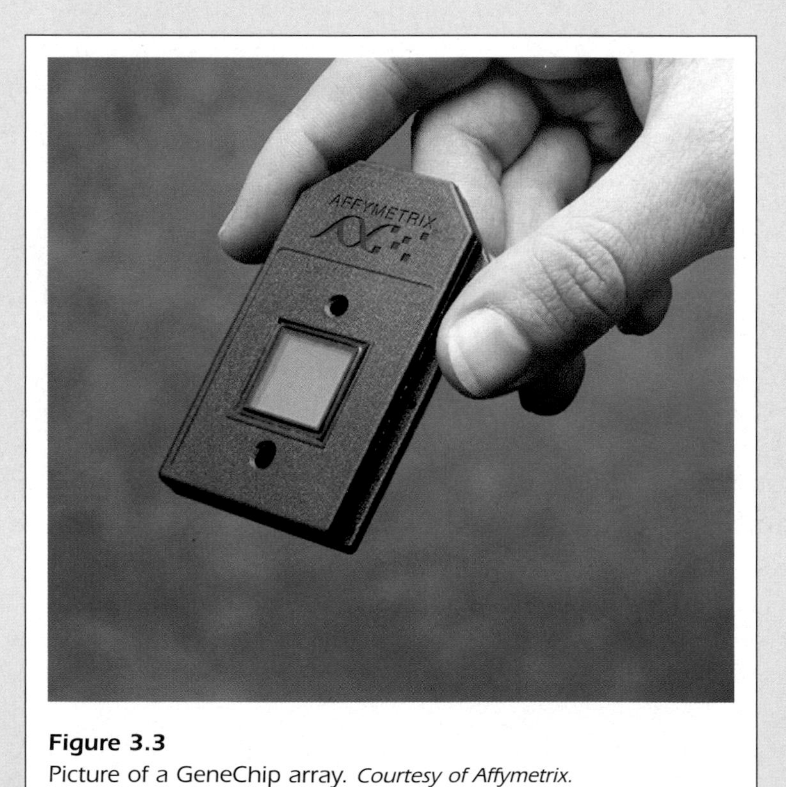

Figure 3.3
Picture of a GeneChip array. *Courtesy of Affymetrix.*

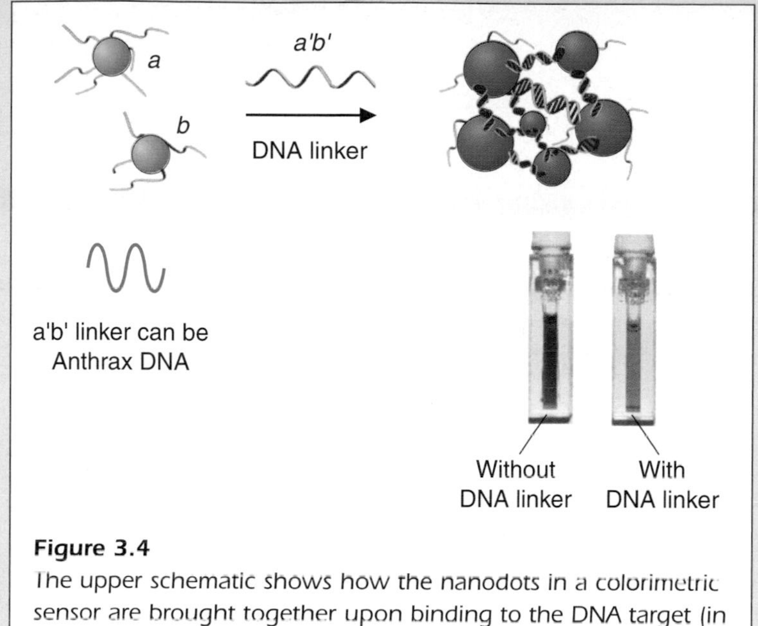

Figure 3.4
The upper schematic shows how the nanodots in a colorimetric sensor are brought together upon binding to the DNA target (in this case anthrax). The clustered dots have a different color than the unclustered ones in the photograph below. *Courtesy of the Mirkin Group, Northwestern University.*

Figure 3.5
This photograph shows pants based on nanotechnology enhanced fabric. The inset shows how the fabric resists the penetration of liquids. *Courtesy of Eddie Bauer.*

Figure 3.6
This photograph shows soldiers using powdered sorbents for decontamination. Newer nanopowdered sorbents will make the process more effective and efficient. *Courtesy of the U.S. Army.*

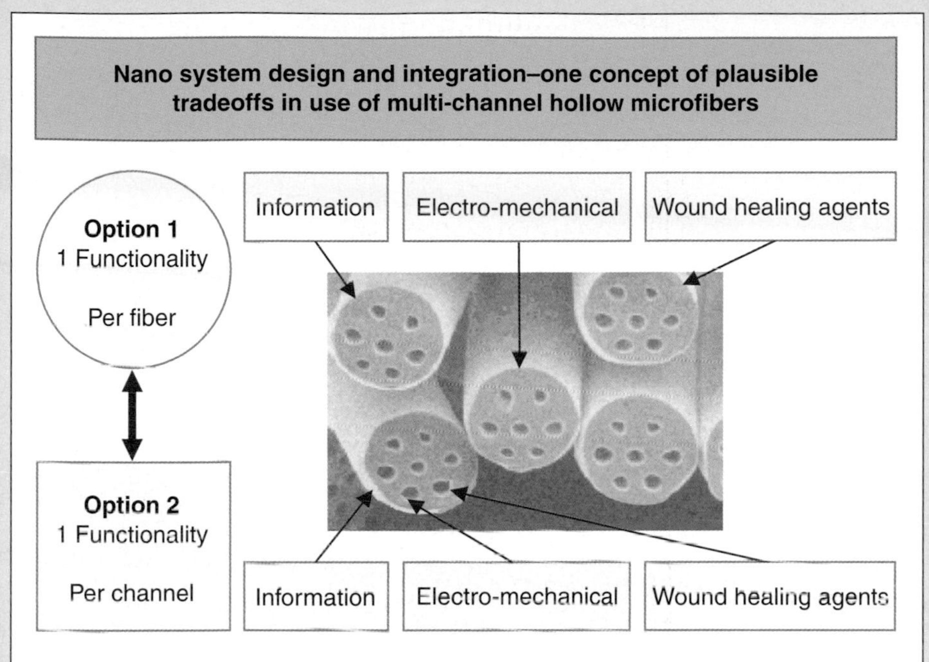

Figure 3.7

A multichanneled fiber like the one shown in the picture might be used to integrate the systems in a soldier's uniform. *Courtesy of DuPont.*

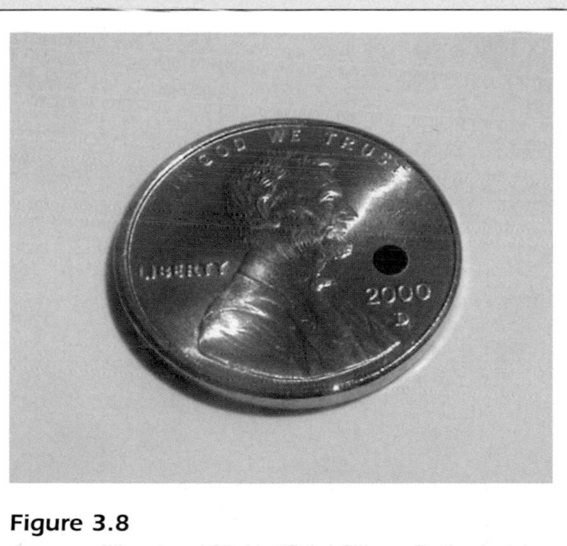

Figure 3.8

A two-millimeter ASR (Artificial Silicon Retina) chip lying on a penny. This chip can be embedded in the eye and may help to restore certain kinds of vision loss. *Courtesy of Optobionics, Corp.*

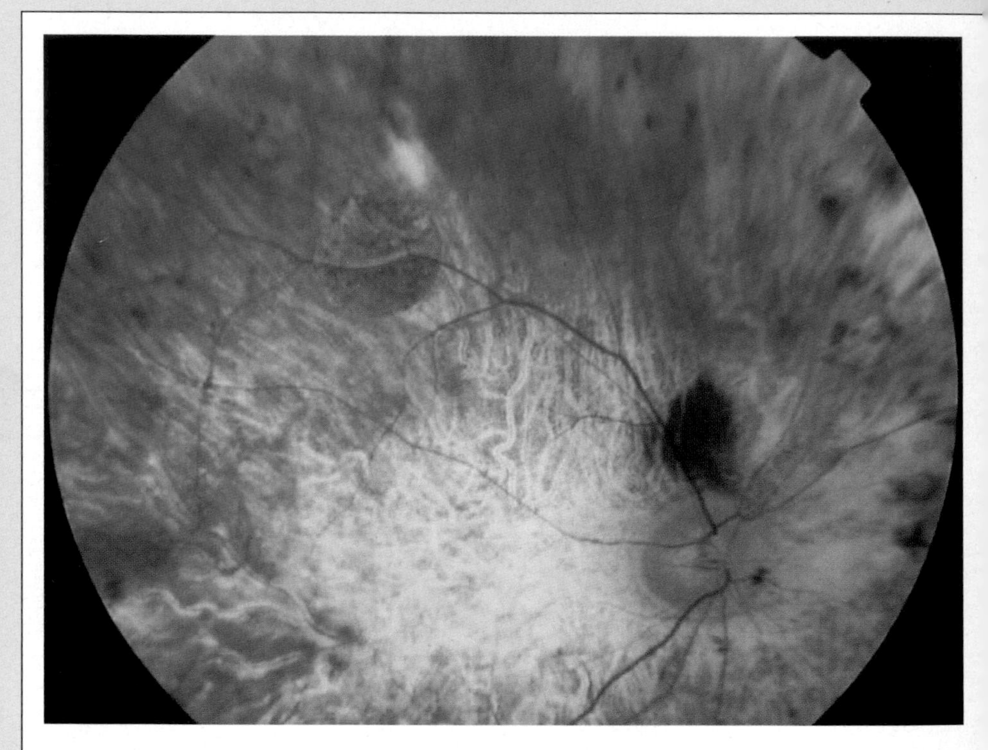

Figure 3.10
The ASR microchip from Figure 3.8 implanted in the human eye. *Courtesy of Optobionics, Corp.*

Figure 3.11
Photograph of the UAV (Unmanned Aerial Vehicle) Predator. *Courtesy of General Atomics Aeronautical Systems, Inc.*

Figure 3.12
A concept unmanned ground combat vehicle. *Courtesy of System Planning Corp.*

Figure 3.13
Photograph of the F117, one of the first stealth fighter aircraft. *Courtesy of Lockheed Martin Aeronautics Co.*

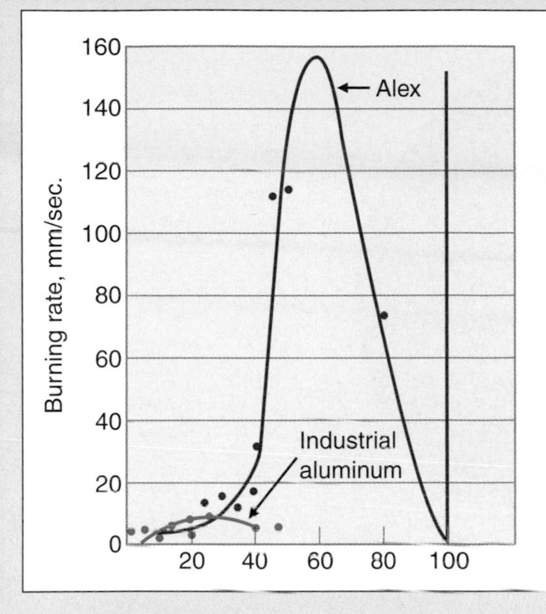

Figure 3.14
This chart shows the relative performance of Alex, a nanopowdered fuel additive, and its microparticle competitors. *Data courtesy of Argonide Corp.*

Figure 4.1
Photograph of a new kind of fire protection glass in action. *Courtesy of the Institute for New Materials.*

Figure 4.2
Photograph showing a lighting reflector that has been exposed to exhaust and other staining. *Courtesy of Fred Simmons, SHC, Inc.*

Figure 4.3
This photograph shows the same lighting reflector in Figure 4.2 after the photocatalytic nanolayers on its surface have been exposed to light and it has cleaned itself. *Courtesy of Fred Simmons, SHC, Inc.*

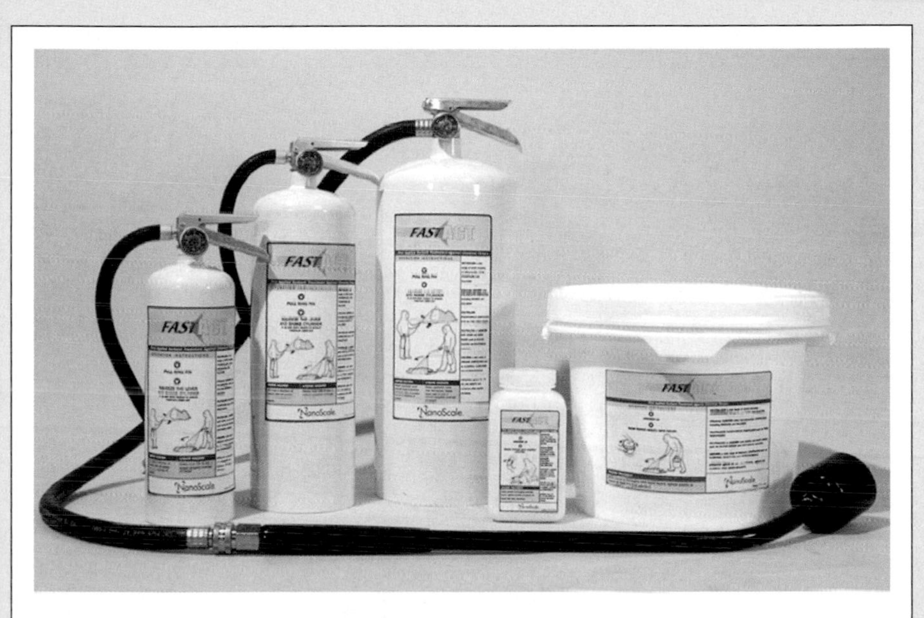

Figure 4.4

This photograph shows delivery systems for FAST-ACT (First Applied Sorbent Treatment Against Chemical Threats), a nanotechnology-based chemical hazard containment and neutralization system which is now available. According to NanoScale Materials, Inc., its manufacturer, FAST-ACT is the safest, most effective dry powder decontamination technology available today. *Courtesy of NanoScale Materials, Inc.*

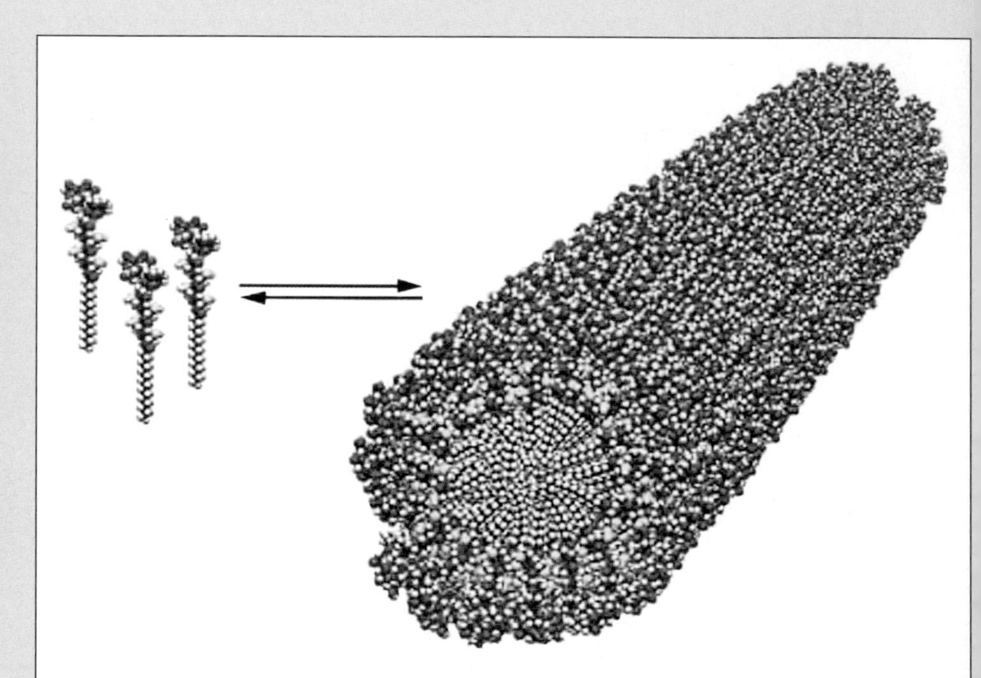

Figure 4.6
Peptide amphiphile molecules, (left) that self-assemble to form a cylindrical mandrel (right) on which bone can grow. *Courtesy of the Stupp group, Northwestern University.*

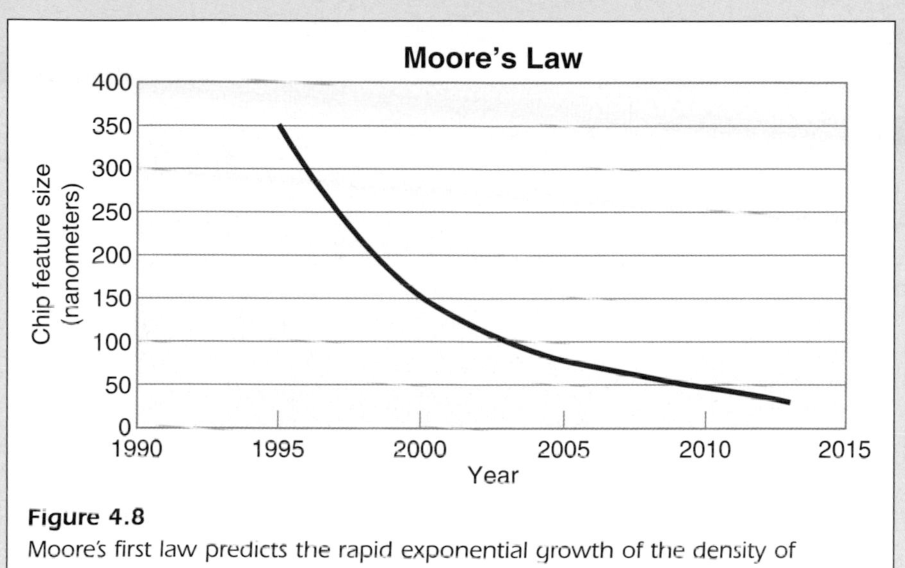

Figure 4.8
Moore's first law predicts the rapid exponential growth of the density of transistors on a computer chip.

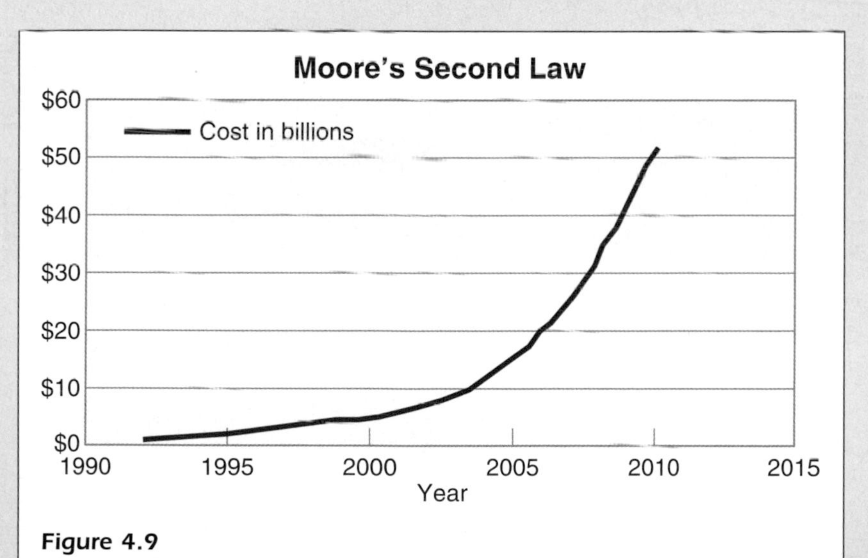

Figure 4.9
Moore's second law notes the rapidly increasing price of fabrication facilities for making computer chips.

Figure 5.2
Hybrid automobiles are powered by both a powerful electric battery and a conventional gasoline engine.

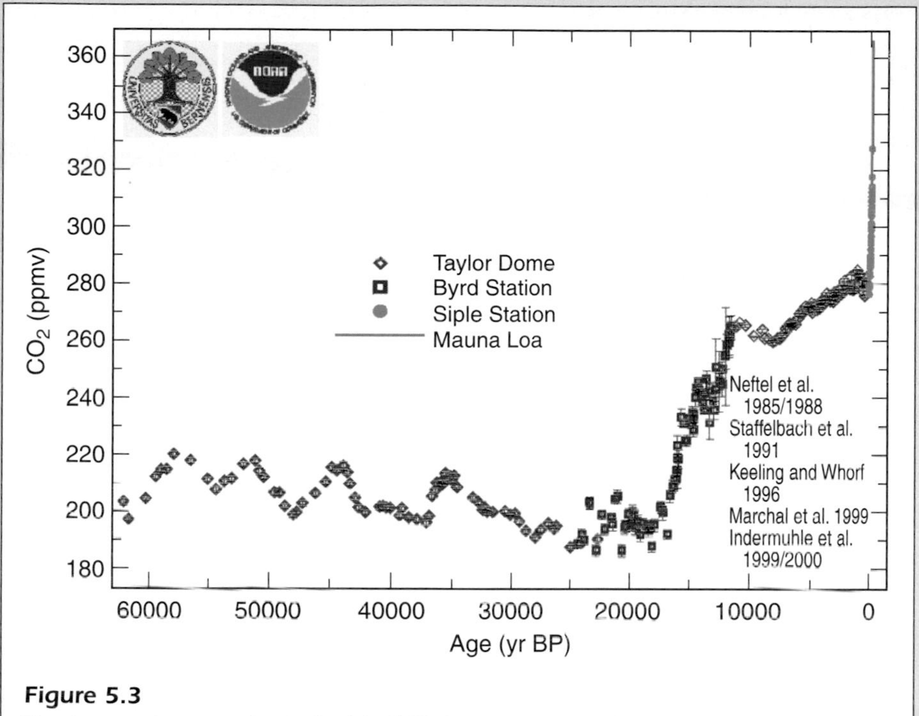

Figure 5.3

The image shows carbon dioxide (CO_2) concentration in the atmosphere over the last 62,000 years. *Courtesy of University of Bern, NOAA-CMDL.*

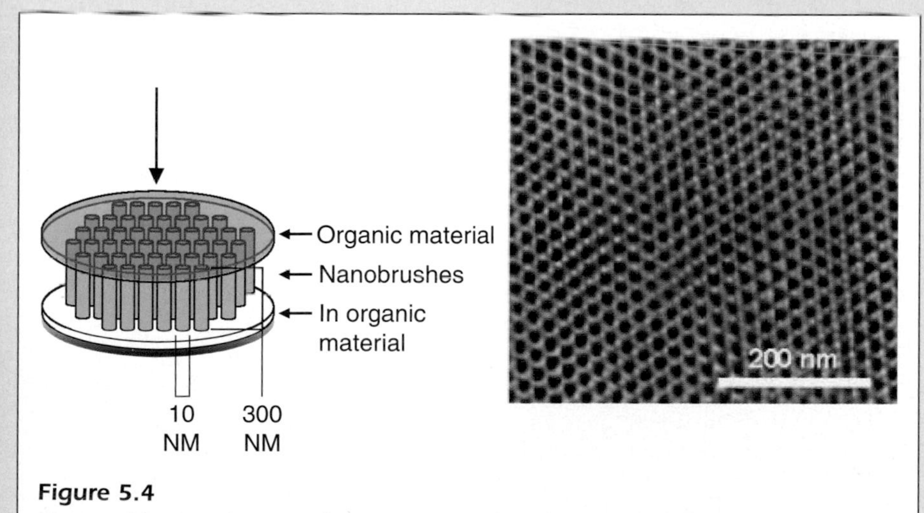

Figure 5.4

Nanoarchitecture image of an enhanced, flexible nanoscale solar energy collector. *Courtesy of Nanosolar, Inc.*

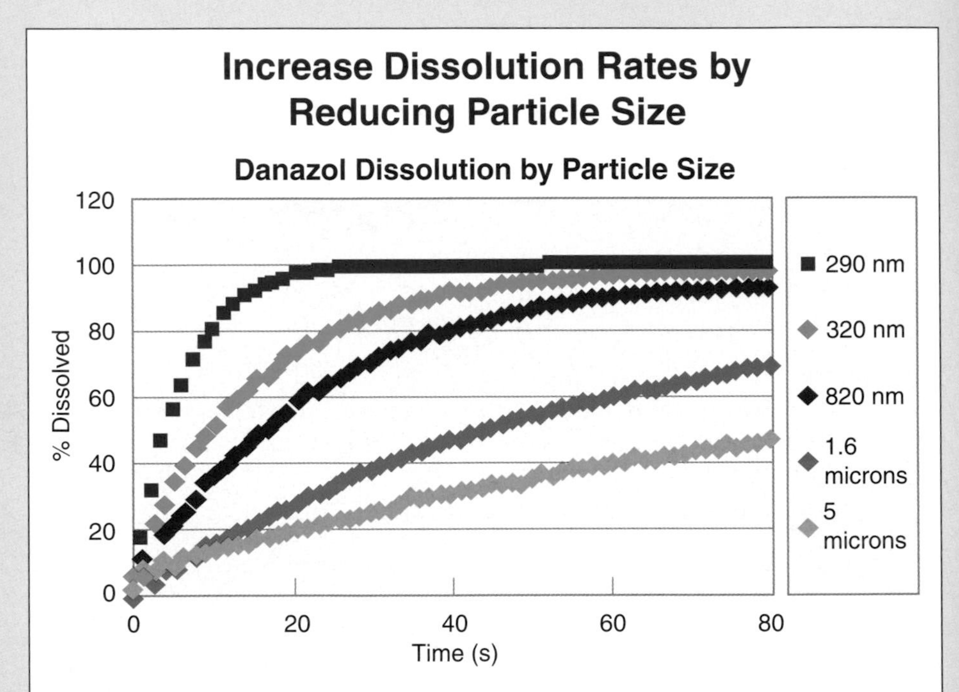

Figure 5.5
Increasing bioavailability (measured by solubility in water) of a medicine upon grinding it to nanoscale size. *Courtesy of Chris Tucker, Dow Chemical Company.*

Decontamination of equipment, buildings, and even people is important enough that it already has a name (Decon) and some component serial numbers (M-11 and M-295) in the U.S. Army. Dual use, both for military and civilian protection, is a clear advantage of these nanotechnologies, because cleaning up the mess is important after an accident such as an oil or chemical spill, just as it is on a battlefield after a chemical or biological attack.

Human Repair

Since nanoscience works close to the interface between man and machine, some terms occasionally slip from one context to another. Machines start to be described in human terms and humans to be described in machine terms. An excellent example of this, and one of the most promising fields of nanoscience, is human repair.

One form of human repair involves fixing "mechanical" problems with the human body, such as bone fractures, torn muscles and ligaments, burns, and cuts by helping the body heal itself. This differs from the conventional method for replacing bones using steel or ceramic implants, stapling tissue, and using other invasive tactics.

Human repair can greatly accelerate the process of healing. Consider the common case of a broken bone. In order to heal, it will probably have to be immobilized. This can be done with a plaster cast for many simple arm and leg fractures, but if the fracture is complex it may involve surgery, bone nails, or even full-body immobilization. See Figure 4.5 for a somewhat grisly example.

It usually takes months for a bone to set and even longer for it to heal completely, but human repair can change this. Research done in Sam Stupp's group at Northwestern Univer-

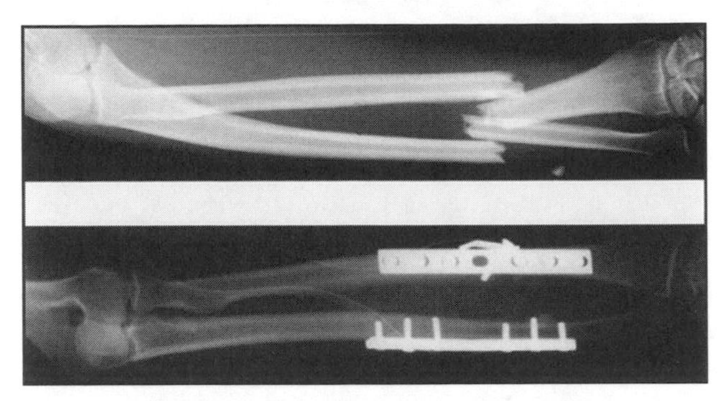

Figure 4.5 X-ray image of a bone fracture and repair. Human repair may make the steel plate and nails shown in the X-ray unnecessary. Courtesy of Dr. Robert Satcher.

sity addresses this very problem. Stupp's approach consists of injecting the site of a fracture with small molecules that assemble themselves into bonelike fibers that bridge a fracture (see Figures 4.6 and 4.7). The self-assembly of these fibers from the liquid can take just seconds, and as soon as it is in place, healing can begin. Osteoblasts, cells that aid in the development or

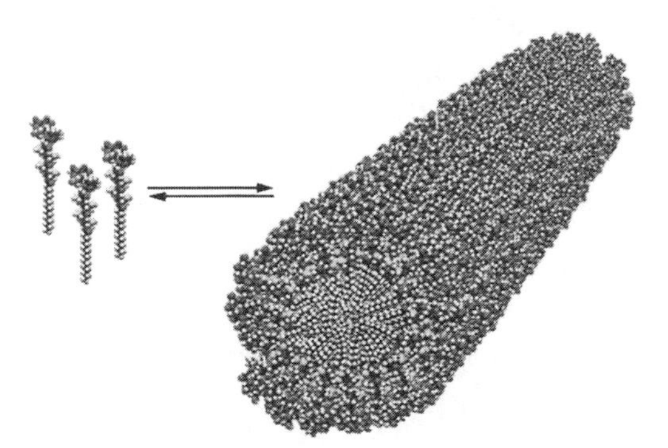

Figure 4.6 Peptide amphiphile molecules (left) that self-assemble to form a cylindrical mandrel (right) on which bone can grow. Courtesy of the Stupp group, Northwestern University.

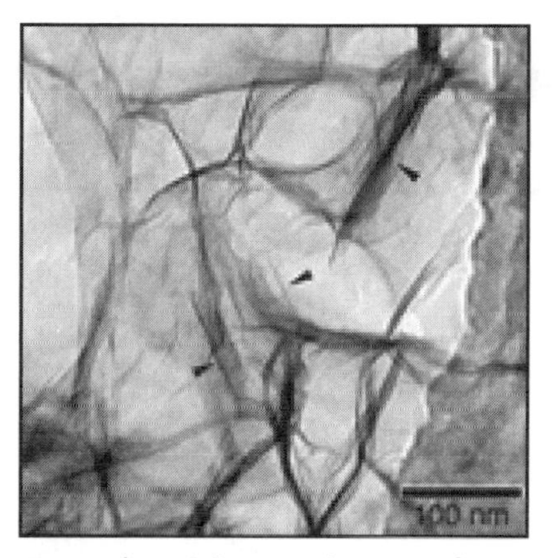

Figure 4.7 Micrograph image of artificial bone fibers. Courtesy of the Stupp group, Northwestern University.

repair of bones, adhere to the nanofibers, and natural bone forms to bridge the fracture.

This approach to human repair can result in dramatically faster recovery from injuries, less surgery, and a better mend when healing is complete. It also does not require steel bolts or implants that must be surgically removed later or that remain with a patient for the rest of his or her life. Almost everyone breaks a bone at some time. Wouldn't it be nice to be able to walk out of a hospital just days after being wheeled in with a broken leg?

Another form of human repair centers on controlling bleeding from traumatic injuries. This innovation will increase our ability to treat mass casualties, and it has obvious dual-use applications.

The Information War

More and more of modern warfare is about who controls information, not just who has the most troops or the best weapons. The information war takes place at many levels, including intelligence efforts before a confrontation, identifying enemy positions, cyber-warfare, and protecting lines of communications. Almost all warfare now has an information component, but we will focus on two or three aspects in which nanotechnology is likely to change the game significantly.

High-Performance Computing

The first and most obvious influence of nanotechnology will be in high-performance computing. It is clear that there is a limit to how much microchips based on current designs can be improved. Already the smallest components are entering the nanoscale, and as they do, engineers are confronting the strange properties of the quantum world. Most projections agree that by the year 2010, Moore's Law, the venerable computer industry maxim that computer components halve in size and double in performance every 18 months, will be at an end (Figure 4.8). To compound this, Moore's less cited second law (Figure 4.9), that chip fabrication facilities double in cost with

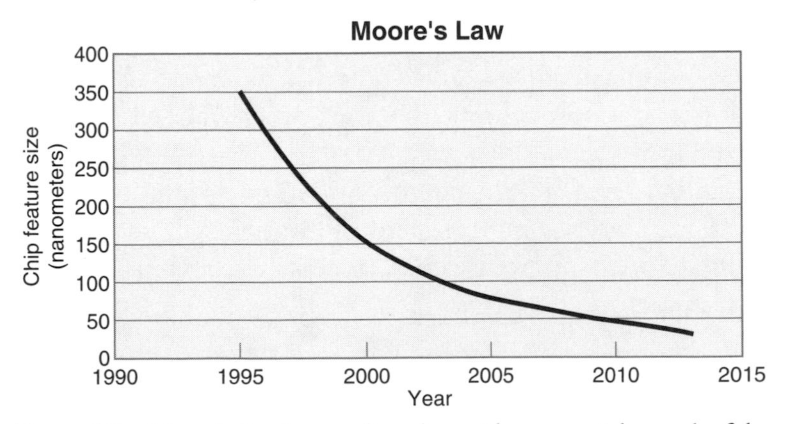

Figure 4.8 Moore's first law predicts the rapid exponential growth of the density of transistors on a computer chip.

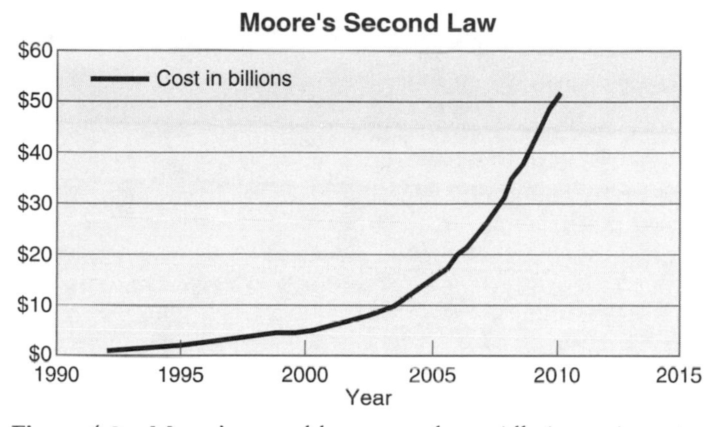

Figure 4.9 Moore's second law notes the rapidly increasing price of fabrication facilities for making computer chips.

each chip generation, shows no chance of abating. This means that by 2010 we might not only start seeing dramatically declining improvements in silicon technology, but that the fabrication lines to manufacture the most cutting-edge chips will cost between $25 and $50 billion. At that price, most semiconductor companies would go out of business.

This is one of the chief drivers that has made nanotechnology so hot. Various new approaches to higher performance computing will only be possible using nanotechnology. DNA computing uses DNA to crunch data as it now stores the formulae of life. Molecular electronics uses individual molecules as electronic components. Bioelectronic computing uses proteins and other biostructures as computer components. All-optical computing uses light instead of electrons for all functions of a computer. Perhaps most powerful of all, the quantum computer can compute problems using the quantum properties of matter. These advances have caused nanotechnology research pioneer Stan Williams of Hewlett Packard to suggest that, in

Kwan S. Kwok
DARPA Program Manager

It is widely accepted that the potential impact of nanotechnology may be larger than that of any scientific field humankind has previously encountered. However, while the development of the nanoscience, on which nanotechnology is based, is important for the initial establishment of a new research area, the ultimate success of the field is measured by societal impact, based on the number and extent of applications resulting from adapting nanotechnology to practice. It is, therefore, of vital importance for the success of this new area to have government-sponsored programs that provide technical challenges to the research and industrial communities so that there is a clear ultimate vision that will result in significant societal impact.

For example, in the area of electronics, researchers have demonstrated the ability to build circuits using individual molecules as circuit elements. In an effort to accelerate the research and development of this technology area, the Defense Advanced Research Projects Agency supports research in nanoelectronics using molecules and other nano-scale components and techniques. The goal of this program is to develop molecular-scale computational systems with logic, and sensory functions that can perform equal to or better than nature's best nose. The planned computational system features complexity comparable to the classic Intel 4004 microprocessor (1971-vintage) but with an area 100,000 times smaller. It will be 100 times as dense, and at least 10 times less costly to manufacture than those presently promised by industry for 2016, at the end of the 2001 International Technology Roadmap for Semiconductors. The sensing structure will feature sensitivity and selectivity capabilities that outperform nature's best nose. The resulting system is scheduled to be completed in 2008 and it is analogous to reducing the current Pentium chip to the size of a pinhead while maintaining its original computational functions. This and other programs could mean 10^{18} bit-operations per second in a volume of about the size of a human head. This goal is expected to be realized in approximately twenty years, in 2024.

However, there is no need to wait until 2024 to see products with "nanotechnology inside." In fact, in a few years it will be possible to develop a memory storage device that holds maps of the world with interactive real-time information and details that are useful for soldiers on the battle-field. This map storage device is analogous to the currently available pocket-size unit that stores 2000 songs, except that the map storage unit with "nanotechnology inside" will be no larger than a coin and will store at least 1,000 times more information than the units currently available.

The twenty-year vision provides an unmistakable direction and long-term challenge for researchers in the nanoelectronics area. At the same time, the immediate targets of building a molecular-scale computational system with logic, and sensory functions that can perform equal to or better than nature's best nose, are clearly established. Visionary input and proper guidance with appropriate challenges established in government programs similar to those given in the example above will very likely shorten the time it would otherwise take for the technology to mature so that practical applications resulting from the new technology can reach the consumer marketplace sooner.

---

spite of forthcoming limitations on current technology, "It should be possible to compute one billion times more efficiently than is currently possible. That means you could hold the power of all earth's present computers in the palm of your hand."

Faster computers are not just essential to the economy. They also allow for a number of security applications. For example, you often show your photo ID to enter a building. Devices exist that can instead scan your hand, eye, or face to determine your identity, check you against a list of restricted people, and provide building security by letting you in or locking you out. These devices are much more secure than a quick inspection of a photo ID, but they have a problem: comparing data from a user to a small library is fairly easy, but if the library of users includes tens of thousands or even millions of people, even the fastest computers can be swamped. The problem is even more severe if a DNA database of known terrorists is to be developed and used to help track them down. Although this database effort is controversial, DNA for forensic use has been common for some years.

Image recognition in satellite photography is another potential nanoapplication, especially since nanotechnology-based computers are showing a particular penchant for pattern recognition. Data storage is yet another—DNA-based storage is one of the most efficient methods for long-term storage where access speed is not an issue, and three-dimensional "holographic" (light-based) storage offers quick retrieval as well as storage capacities much greater than anything we have today. Both of these data storage techniques exist already, though they are not yet commercially available. Nanotechnology has already made gigabyte hard drives reality, and that is proving to be a tiny fraction of what the tiny can do.

The loftiest goal of high-performance computing remains artificial intelligence (AI). AI would allow robots to take on many of the most risky jobs, such as serving in front-line ground forces. True, self-aware AI is probably still quite far in the future, but machines that can think their way out of complex situations may be just around the corner.

Cryptography

Code-breaking is essential to any serious intelligence gathering. If a nation wants to intercept terrorist communications or enemy transmissions in the battlefield, it must have good code-breaking ability. Arguably the tide of World War II turned for the Allies when Alan Turing and his colleagues broke the Nazi cipher called Enigma. By 1942 the allies were decoding tens of thousands of Nazi communications each month, and the intelligence they accumulated has been estimated to have shortened the war by at least two years (though it is hard to know what effect the atomic bomb would have had). However, Turing's efforts took several years, an amount of time which is impractical in modern warfare and intelligence gathering.

Today, advanced cryptography is widely available. Keys and encryption schemes can be changed quickly. Terrorists and hostile military forces take advantage of these technologies to make it difficult to recover information from captured computers or intercepted transmissions.

Many modern digital encryption schemes (such as RSA, a common scheme for e-commerce applications) rely for their security on the idea that large numbers are very difficult to factor and that the time it takes to factor them goes up exponentially with the size of the number. While a computer can in principle break these codes using a guess and check technique, breaking today's best commercially available codes using this approach would take a modern PC (circa 2004) more time than has passed since the fall of Rome. For this reason, the common commercial encryption schemes that permit data security within our society can also allow terrorists and other criminals to exchange data with almost perfect security.

Quantum computing is one solution that could turn the mathematics of code-breaking on its head. A quantum computer can, effectively, work on all the possible solutions to the problem at the same time rather than in sequence. In reality the solution is a bit more complicated than that, but there are reasons to believe that a single quantum computer could solve an encryption problem far beyond the reach of the mightiest PC—and it could do so in just a fraction of a second.

Breaking codes is more of a double-edged sword than most military or defense technology. The overwhelming majority of people using cryptography are employing it for lawful and decent reasons including trading stocks, buying goods, or protecting the privacy of their email over public networks. Even more than most nanotechnology, decryption technology with this much power must be used sparingly and its abuse must be punished diligently. We will talk more about this in Chapter 6.

Quantum computing has applications far beyond decryption. Searching massive amounts of data and even modeling other quantum mechanical systems are two obvious examples.

Flexible and Pervasive Computing

Relatively little thought has gone into the form, shape, or mechanical properties of modern computers. One or more silicon chips form the core of almost all electronics from computers to music players, and the silicon chips are still centimeter-scale pieces of silicon in plastic packages often trimmed with metal heat sinks. The chips sit on circuit boards of rigid plastic. This is a perfectly good way to build electronics for many applications, but it prevents them from being integrated into clothing, packaging materials, wall paint, and many other places where they could be useful.

It may not be obvious why it is desirable to put computers into wall paint. Many people justifiably feel that computers are already too much a part of their lives, but the applications for tiny, flexible computers (so-called "pervasive computing") are very intriguing. For example, pervasive computing could have household applications, such as controlling lighting, temperature, or music volume based on inputs ranging from what you are doing to the expression on your face.

In addition, pervasive computers have security applications when coupled with appropriate sensors, since they cannot be disabled easily the way motion detectors can. Unlike most security systems, which operate best after business hours, they can also allow the continuous monitoring of an area even during periods of normal activity. Finally, they are necessary for many of the soldier applications we've discussed, including the next-generation uniform.

So far the most promising avenue for this kind of capability is a nanotechnology called molecular electronics. Molecular electronics involves using individual molecules as components in electronic circuits. This is the ultimate level of miniaturization for electronics of the type that we are used to—transistors, capacitors, and the like. In addition to being very small, molecules (as opposed to metals or crystalline materials like silicon) are generally fairly soft and flexible. Molecular components may not need to be affixed to a rigid plastic board the way current electronics do. Instead, it may be possible to put them on softer polymer surfaces or other flexible backplanes. This would allow them to be used for many kinds of computing applications.

Molecular electronics will initially be soft and cheap. For example, molecular thin film transistors (and we mean *thin*—just a few nanometers in thickness) now being developed by Phillips, Lucent, and others, are designed primarily for disposable applications like identification tags to label soup cans, pieces of mail, or even individual pills. The soup and pill labels would help track possible contaminations, including terrorist modifications, and allow fast recall. Recall the concerns with meat product contamination and cyanide in Tylenol bottles.

System Diversity and Survivability

One of the greatest threats to American security is the vulnerability of our computing and communications infrastructure. This comes in several forms. The fact that a single computer architecture (the PC) and a single operating system family (Windows) now comprise the overwhelming majority of our data systems represents a huge threat, since any weakness discovered in either design may be exploited on millions of other systems. This problem is similar to dependence on a single crop for food. For example, in the Irish potato famine a blight wiped out the whole crop of potatoes and there was no alternate crop, such as corn or wheat, to see the population through the crisis. High-impact computer viruses have shown that the analogy holds for computer systems. What often starts as a prank can result in billions of dollars in damage and lost productivity. While nanotechnology is not currently poised to do anything about operating systems directly, it may help to evolve the underlying architecture of systems, diversify them, and make them more secure.

As mentioned earlier, nanotechnology is already promising at least five possible major revisions of traditional computer architecture. These include all-optical computing, bio-electronic computing, DNA computing, quantum computing, and molecular electronics. Some of these, like all-optical computing and molecular electronics, are general purpose and may work quite similarly to current machines. Others, like DNA and quantum computing, will be much more effective for speeding certain kinds of computations (such as code breaking) and can offer some types of tamper-free computing, but may never find application in general-purpose computing. All of them change some basic rules.

While we've talked a lot about chemical and biological weapons, we have not spent much time discussing the third major category of weapons of mass destruction—nuclear weapons. There are many reasons for this. Not even nanotechnology will protect a city from a megaton nuclear device. It may be helpful in detecting the weapon or even shooting down a missile carrying one, but there are no technologies on the near-term horizon which will do much to protect us if the worst happens.

Having said that, a nuclear weapon from a less-developed state is unlikely to be an advanced hydrogen bomb with a megaton yield. Smaller, so-called "tactical" or "dirty" nuclear weapons are much easier to make, buy, or smuggle into a target zone. They cause a great deal of immediate damage, but there are other environmental effects of a nuclear detonation, which must be dealt with similarly to remediation of chemical and biological weapons. Another effect of such weapons is the nuclear electromagnetic pulse (EMP). An EMP is a wave of massive energy that can disable or destroy electronics well beyond the reach of the actual nuclear blast.

The military specification for electronics requires ruggedness well beyond most civilian specifications. Not only must military electronics operate in hotter, colder, and damper climates than civilian electronics, they must also have some resistance to EMPs. While it is possible to improve the resistance of electronics to EMPs, it is very difficult to shield them completely, and most shielding requires significant sacrifices in terms of weight and performance. There is reason to believe that all-optical computing and DNA computing as well as some other nanotechnology-based computing options will be naturally resistant to EMPs. They will also have better performance in other hostile conditions including outer space.

All-optical computing (using light rather than electricity to process data) is one example that will not only resist EMPs, but will also be less sensitive to electronic eavesdropping. Since the optical signal is confined to a fiber much more tightly than an electronic signal is truly confined to a wire, it is harder for an external piece of equipment to listen in. The limiting factor to creating an all-optical computer is the design of a small, efficient optical switch to perform the same duty as a transistor.

Bulkier optical switches for all-optical communications networks exist, but optimization to the point of usefulness for a complete computing platform has not yet been done. It will be done eventually, and all of the most promising designs to date are based in nanotechnology.

All of the nanotechnologies that we have just surveyed for homeland security exist in some form. Some are already commercial products and some are still several years from development. Even the far-out notions such as quantum computing are close enough to hand that corporations, venture capitalists, and universities are becoming interested. For example, D-Wave, a Canadian startup interested in practical quantum computing, raised a large round of private equity under the most difficult conditions at the beginning of 2003.

There are dozens of other ways in which nanotechnology could affect homeland security, and many of them will only be recognized as fundamental research continues. Unlike defense technology, however, homeland security technologies will impact all of our lives directly. Some, such as human repair, will only change life for the good. Others, such as advanced cryptographic technology, could be much more threatening. And still others, like AI or pervasive computing, could change our day-to-day lives as much in the next 15 years as electronics, the Internet, and telecommunications have done in the last 50.

Environmental and Economic Aspects of Nanoscience

"Nature never did betray the heart that loved her"
—William Wordsworth

It is important to remember that both security and nanotechnology are complex subjects. The threats to societal security come in many guises, from something as simple (if frightening) as a bomb attack on an office building to more subtle means such as disseminating computer viruses or electronic disinformation or tampering with drug packaging and distribution.

Security concerns go well beyond those associated with terrorism, ranging from new diseases like AIDS, SARS, and Ebola to the tremendous challenges associated with climate change, energy supplies, and global warming. Nanotechnology can be of great value in addressing some of these concerns, but others extend into realms beyond the scope of physical technologies.

To take one simple example: If there were no motivation for acts of terrorism, there would be no acts of terror. These motivations can come from economic inequity and uncertainty, from religious belief or incitement, from ethnic or religious hatreds and traditions. They may even come from a combination of factors featuring civic pride—every Super Bowl win or NBA championship might well start looting, rioting, and burning in the winner's city.

In the last two chapters of this book, we address that part of these motivational aspects that arises from concerns about the development of technologies. All technology (including nanotechnology) can change the economic, education, environmental, and employment patterns in society. This change can cause concerns about everything from globalization to job security to environmental dangers. These concerns extend well beyond the scope of our discussions, but are a major aspect of terrorism and of homeland defense.

Fabrication

Nearly all manufacturing processes have substantial environmental impact. At the height of the industrial revolution in England, streams were polluted with animal wastes and chemical byproducts, and the skies were often thick with killer smogs of particulate carbon smoke and sulfur dioxide. In Silicon Valley, several of the byproducts of high tech fabrication have created large environmental problems. Almost no American city has drinking supplies that are clean enough to use without treatment, and brownfield and Superfund sites (industrial areas that were left polluted by a previous owner) are becoming more and more abundant. Anyone who has lived near a steel mill, oil well, tannery, foundry, or paper plant can vouch for the unpleasant environmental aspects of manufacturing.

There is a great deal of concern that nanoscale manufacturing could augment these problems. The reality is probably just the opposite. Fabrication of nanostructures is in many ways less hazardous than manufacture of more traditional industrial products because the control must necessarily be much better.

In an automotive assembly line, for example, welding is often done in the open and little can be done to contain waste gases, so these are simply vented. Raw materials for macroscale manufacturing often include nasty agents like chlorine or the hydrogen cyanide (HCN) gas we talked about as a chemical weapon. In the manufacture of nanostructures, however, not only are the amounts of material used typically smaller and the components more benign, but the precision and chemical sensitivity of the manufacturing processes require that fabrication be done in closely controlled systems under laboratory conditions. This makes waste easier to capture and, once captured, easier to treat (Figure 5.1).

Still, nanoscale manufacturing involves dealing with materials that could have harmful effects very different from those

Figure 5.1 Nanotechnology fabrication facilities are likely to be enclosed and easy to prevent waste from seeping into the environment. Nanosolar's fabrication facility is one such example. Courtesy of Nanosolar, Inc.

used in traditional manufacturing. Consider biological nanostructures, for example. The design of sensors, improved drugs, and a variety of nanotechnology-powered devices involves using DNA, bacteria, and viruses. Some people worry that biologically active agents might be released into the environment where they could interact with other biostructures (including plants, animals, or people), perhaps leading to substantial new health hazards. This concern is similar to the worry about genetically modified organisms in food, or (in an earlier time) concerns about medical waste and hybrid farm crops including corn and wheat.

With both biological structures and their nonbiological counterparts, the fundamental issue in limiting environmental damage is controlling the effluent streams from manufacturing or health facilities. In a fabrication plant for biological nanostructures, this task should be fairly straightforward, since most operations would probably be carried out by robots in controlled environment chambers (analogous to a semiconductor plant's clean room). Air coming into and out of the chamber can be filtered, purged, and irradiated so that no biologically viable products are released into the environment. The few workers who do enter can wear nano-enhanced defense suits, and outside surfaces of chambers and equipment could have biocidal coatings.

Of course it is true that these filtration systems might break down, but because the volumes involved would be relatively small and the manufacturing processes very highly controlled, the rate of emission of viable biological material into the environment could be made essentially negligible. Even if such materials were to escape, nearly all of them would break down rapidly—both simple DNA strands and proteins denature quickly when exposed to the outside environment. Even if the scale of manufacture or the extent to which people are directly involved is greater than expected, there are established procedures for running these types of plants safely. After all, the problem of containing and treating biological entities used in manufacturing is solved every day in the world's pharmaceutical factories.

Another concern raised by activist organizations like the ETC Group is that even nonbiological nanostructures (which one might expect to be benign) can in fact be highly toxic. There are a few reasons for this. One is the extreme reactivity of nanoscale materials (the same property that makes them so good as catalysts), and the other is that these ultra-fine particles are the right size to be absorbed by cells and membranes in much the same way that asbestos is. While sheets of asbestos or carbon are not particularly toxic, tiny airborne particles of either could be trapped in the respiratory tract and become extremely carcinogenic. In the manufacturing context, this problem can easily be addressed using the same techniques we've just discussed. However, more work is required to determine how great these risks really are, especially for manufacturing such products as skin lotion, medicine, or other material to be used on people. Absorption into the body is a potential issue only for certain classes of nanostructures. Integrated nanodevices, where a nanostructure is physically integrated into a larger device such as a chip, smart materials, and most other developments from nanotechnology, will be immune.

In sum, the fabrication of nanostructures has some unique challenges, but nanoscience also provides solutions to most of them. Additional regulatory oversight for nanomanufacturers is necessary (we'll discuss this in Chapter 6), but on balance nanofabrication should produce less waste and few new environmental hazards if it is well managed. Such management is well worth the effort, since the next few decades have the potential to be the best postindustrial period yet for the Earth itself and for the people on it.

Remediation of Ongoing Environmental Issues

Environmental awareness has been one of the most striking, impressive, and important political changes in the past half century. Although environmental activism has roots going back to Thoreau and beyond, the naturalists, writers, sportsmen, artists, political figures, and social leaders who brought about the modern environmental awareness revolution have contributed tremendously to society's evaluation of itself and perhaps to its very survival. Their contributions create the possibility for living in a friendlier, safer, more beautiful, and more rewarding world.

While concern about the environment is abundantly justified, nanoscience challenges the common activist argument that industrial and economic growth are necessarily incompatible with sound environmental management. Many ongoing environmental issues have been caused by development and industrial progress without appropriate environmental considerations, but nanotechnology could offer cleaner ways to make things, as well as substantial help in remediating existing environmental hazards.

The three main areas of environmental concern are air, earth, and water. Both air and water can be filtered to remove toxins and pollutants in much the same ways that we looked at in Chapters 3 and 4. Because many common pollutants are composed of particles that are themselves nanosized, nanoscale filters are ideal for purification. Such ultra-filtration methods are already common in industrial areas, where they have proven to be relatively cheap and very effective, but new nanotechnological developments should permit these filters to remove many new types of waste products and to be longer lived, many times more efficient, and capable of handling much larger volume. It is difficult to imagine cleaning a large polluted lake or a city air shed by filtration—how do you get that much air or water through a filter? However, filtration or treatment closer to the source, as with catalytic converters on cars and sewage treatment facilities, can result in the reduction or elimination of pollutants before they ever get into the environment. If these devices are adopted (and incentives or regulation may be required for this to happen), large-scale postpollution remediation will not be as necessary in the future as it has been in the past. Once the rate of new pollution diminishes, the Earth has shown that it has a wonderful ability to heal itself. Many of the processes it uses involve natural nanotechnology.

In some cases, though, the natural rate at which the earth heals itself is not sufficient. The Valdez oil spill and a chemical weapons attack are obvious examples. In these cases a more aggressive approach to remediation may be necessary. Although there is no universal solution, nanotechnology is providing approaches for dealing with some of the more common and aggressive pollutants. For example, particular ions such as chromate or arsenate can end up in water supplies and are extremely carcinogenic. Ions cannot be easily filtered, but a simple chemical exchange can replace these toxic ions with more benign alternatives such as ferrous iron. While this has been known for some time, the efficiency of this replacement process is limited by the amount of surface area of the reactants. As the particles of treatment chemicals are reduced to the nanoscale, they essentially become all surface. This means that the process of replacing toxic ions with benign ones can be greatly sped up, made much more efficient, and made much cheaper.

The capability of nanostructures to sense certain molecules could also be useful for dealing with a number of important environmental problems, especially those that result from military campaigns. For example, one of the nastiest residues of military conflict is unexploded ordnance, including bombs, grenades, and landmines that can remain hidden on the battlefield and cause danger, fear, injury, and death long after the war is over. Sometimes this ordnance can be destroyed using bulldozers and earthmovers, but rocky or mountainous terrain can make that virtually impossible. Even human mine clearing is not always 100% effective (though it is nearly 100% risky), and in places like the Middle East there are large tracts of land that are only approved for animal pasturage since no amount of clearing can guarantee that all the mines are gone. This isn't a problem limited to the developing world. There are still unexploded shells from World War II on Salisbury Plain in England and in Danish coastal waters.

Until recently, soldiers looked for mines using metal detectors. Military technology kept pace, however, and mines now exist that are mostly plastic and thus virtually impossible to find with a metal detector. Chemical sensing provides an alternative. Since nearly all unexploded ordnance contains nitrate propellants and since nitrates are relatively easily sensed (the white fabrics used in airports to swab luggage do just this), nanoscale sensors may be able to find this unexploded ordnance even if it is buried deep in the ground. Removing the scourge of mines and bombs from the fields in which they are strewn would be of tremendous value in returning the earth to its productive and rich self. Finding hidden weapons is also crucial for dismantling the military apparatus of a foe once a battle is over, and it would be a boon to weapons inspectors trying to prevent a war from happening in the first place.

Public health depends upon the clean environment, but certain structures, notably microorganisms, can affect health even in very small amounts. Nanoscience, through the development of integrated sensors, offers the possibility of total monitoring for public health applications. Such total monitoring is not the invasion of personal privacy as implied by its name. Rather, it means that air, water, and soil could be continuously monitored for the appearance of any known toxic agents. Such monitoring could allow instantaneous response to chemical leaks, filter failures, industrial accidents, and, of course, bio- or eco-terrorism.

A combination of these techniques—sensing, nanocatalysis, ion replacement, nanofiltration, pervasive computing, and others yet undeveloped—can mightily increase the power of the tools we have to fight pollution and clean up our environment. Some of them are in use already. We hope more will be deployed in the next few years.

Transportation

Transportation is one of the major polluters of the air. Estimates for the total contribution of transportation to the air pollution burden in the United States and other developed countries run up to 50%. The results of this air pollution are obvious in cities such as Los Angeles and Atlanta. Replacing diesel or internal combustion engines with portable electric power provided by fuel cells or batteries would cut the emissions of carbon monoxide, particulates, nitric oxides, and sulfur oxides to essentially zero for all cars, trucks, and trains powered by these alternative means.

The idea of using electric powered cars goes back to the beginning of the century (some readers may remember Grandma Duck's Detroit Electric, built in Michigan in 1908), but as of 2004 there are still very few electric powered vehicles on the roads. There are a few reasons for this. Early electric vehicles could not match the acceleration or performance of their fossil-fueled cousins, regulatory pressure such as California's zero emission plan and federal clean air guidelines have been scrapped under pressure from corporate interests, and not enough research has been done to make fuel cells and batteries efficient enough to compete on their own.

Fortunately, the situation is beginning to change. Fuel-conscious buyers ranging from religious groups to corporate fleets are pressuring manufacturers for ecologically responsible vehicles. Also, slowing sales of conventional vehicles are forcing automakers to look for the next big thing. Investment in fuel cell research in particular has soared in 2003. Hybrid vehicles (see Figure 5.2), which use a combination of fossil fuel (for rapid acceleration) and electric power, are gaining some popularity already, and they are a very attractive medium-term solution. Nanotechnology is promising to improve both batteries and fuel

Figure 5.2 Hybrid automobiles are powered by both a powerful electric battery and a conventional gasoline engine.

cells to the point where they may become the sole power source for a car or truck without huge performance sacrifices.

Improving batteries and fuel cells can be done in a number of ways. The key is to improve the power source by making the chemical reactions at the electrodes and the motion of ions within the electrolytes faster and more efficient. This translates both into higher energy density (energy to weight ratio) for longer running time between recharges or refueling and into higher power for better acceleration. Interestingly, some of the most exciting fuel cell research is happening in Texas despite that state's considerable stake in the fossil fuel economy.

Efficient fuel cells for other purposes already exist. NEC has announced a laptop computer based on a methanol fuel cell that it claims can run for five hours, and a 40-hour version should ship by 2005. This means that these fuel cells will have ten times the performance of the best lithium ion batteries—and they're just getting started. The fuel cell industry is still in its infancy, and when it matures to the point that batteries have, the results will be staggering.

Fuel cells work by extracting energy from a chemical reaction rather than by burning fuel like an internal-combustion engine or storing energy like a battery. This means that spent reactants can be captured within the cell rather than vented off (as in the exhaust system of a car) so they can be disposed of cleanly. While different designs are necessary to use different kinds of fuel, successful fuel cells have already been run on a variety of fuels including methanol, pure hydrogen, and methane. Methanol is a common additive in gasohol and other fuels since it is relatively easy to make from natural gas, compost, or other waste products. This makes it a very attractive candidate in the short term, though other fuels have better long-term potential. One of the other options, natural gas or methane, is a common waste gas from landfills. (The fires you see at night on most landfills are caused by vented methane being burned off.) The use of methane means that fuel cells could be installed at landfill sites, where they would be powered by products normally considered to be waste.

While batteries don't generate energy the same way fuel cells do, they allow us to use electricity generated at industrial power stations, which are much more efficient in converting fossil fuels to energy than current cars are. Batteries also eliminate the need for a gasoline distribution system. Using either of these alternative energy sources would permit great reduction in air pollution load directly from transportation.

Energy

The supply of energy is probably the most important physical problem that civilization now faces. As Richard Smalley has pointed out, there is no driving economic or environmental problem on Earth that will not be alleviated or at least reduced if a source of readily available, clean, and inexpensive energy could be developed. Energy powers industry, heats homes, and moves goods. It can also be used to separate salt from seawater, to grow crops, and to eliminate drought and hunger.

Richard Smalley
Gene and Norman Hackerman Professor of Chemistry
Professor of Physics & Astronomy
Rice University
1996 Nobel Prize Winner

The problem of energy has vast political, economic, and social aspects that have been at the root of most wars and much of the political strife of the last century. Still, at its root, energy is a technical problem, and it will have technical solutions. We just need to find them.

The new energy technology must not involve burning carbon. When all the costs are figured in, fossil energy is too expensive now, and can only become more expensive in the future as resources are depleted and billions of people in developing countries begin to live modern lifestyles. The consequences of burning all this carbon and discharging vast amounts of carbon dioxide in the atmosphere are now becoming clear. We must find another way.It can only be nuclear. With our current understanding of physics, only nuclear fission or fusion can produce energy in the vast amounts we need.

If the answer has to be nuclear, it would be best to be far, far away from the reactor. Nature has given us one such reactor, and provided the necessary distance and shielding. It is our sun.

There is plenty of energy from this natural fusion reactor to provide all our energy needs for centuries to come. We just don't know how to harvest it, store it, transport it, and use it in the amounts we need. The technology that will do what we need does not yet exist. It will come from discoveries in basic science, particularly in nanotechnology.

I am confident we can solve this problem of energy supply. But it will take revolutionary breakthroughs in physical sciences and engineering, fields that we haven't been emphasizing enough in the past couple of decades, particularly in the United States. And when you look at the sort of breakthroughs that are going to be necessary, you'll realize that most of them fit in the broad definition of nanotechnology. In order to make these breakthroughs happen, we will need to take this problem much more seriously than we have so far for any project we've confronted in this country since the Apollo project.

As a result of the program addressing this problem, American youth will enter the physical sciences and engineering in a way that they haven't pretty much in the history of the country. They will be inspired not so much by the notion that they're going to get rich, but by the inner sense of idealism that is so prevalent in youth; and their sense of mission. American boys and girls will have this sense of mission because they know that their generation is the one that has to solve this problem. I can tell you from the young kids that I've talked to that this sense of mission is strong. You can hear a special timbre in their voices when they talk about it.

Clean energy would have an overwhelmingly positive effect on the environment. It could reduce, eliminate, or potentially reverse global warming and climate change. These are the two great environmental time bombs. If they are not stopped, they will cause great harm to our society, resulting in everything from flooded coastlines and melted icecaps to changed rainfall patterns that produce deserts where farms once existed. As shown in Figure 5.3, the amount of carbon dioxide (one of the important greenhouse gases that causes global warming) in the atmosphere has gone up more in the last hundred years than in the previous half million years. The reason for this is simply more human life on the planet leading to more industrial production, more breathing, more burning, and thus more carbon dioxide.

The solution to global warming, as was recognized in the Rio and Kyoto Agreements, lies principally in the reduction of carbon dioxide emission into the atmosphere. Almost all carbon

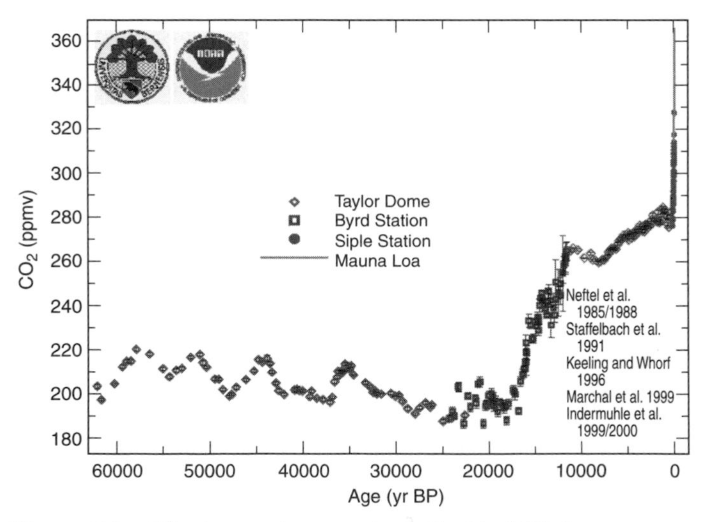

Figure 5.3 The image shows carbon dioxide (CO_2) concentration in the atmosphere over the last 62,000 years. Courtesy of University of Bern, NOAA-CMDL.

dioxide is produced by burning carbon-containing fuels. Animals breathe out carbon dioxide because they burn sugars that contain carbon (though respiration constitutes a tiny fraction of carbon dioxide emission worldwide). Fires send out carbon dioxide because we burn wood, coal, oil, peat, or paper, all of which contain carbon. Greenhouse gases like carbon dioxide are a threat because when they are released into the atmosphere they block the escape of the heat caused by sunlight shining on Earth, effectively making the earth into a greenhouse or reflector oven. This is the major source of global warming, and one that we will have to control if society as we know it is to survive.

Dealing with the carbon dioxide load and global warming will require the development of energy sources that do not involve combustion of hydrocarbons. Petroleum should be used mostly for its capability to make materials such as plastics or pigments, rather than for burning. This would also have substantial strategic advantages since it would make the United States independent of foreign oil. Domestic petroleum supplies are more than ample for nonfuel purposes.

Carbon-free sources of energy include wind, geothermal, tidal, nuclear fission, nuclear fusion, and solar energy. While conservation is not strictly a source of energy, more efficient use of existing energy resources and reduction of unnecessary use also help relieve the strain. All of these avenues are important, and all of them are being pursued. Nanoscience and nanotechnology can help substantially in all of them, but perhaps most obviously in conservation, in the development of solar energy, and in improving the efficiency of existing devices.

Let's start by looking at solar energy. If we could design nanostructures capable of capturing sunlight with an efficiency of 50 percent, then the entire amount of energy required to run every home, business, car, truck, and toy in the United States could be captured by erecting solar energy cells on only a 40 mile by 40 mile area of land in the Nevada, Arizona, or New Mexico desert (based on projections from the National Renewable Energy Laboratory [NREL]). Of course this energy would have to be shipped and stored, but nanoscience (and other scientific approaches) can allow us to capture the energy that the sun sends to Earth every day, energy that is not now used.

Essentially all the energy resources on Earth originally came from the sun. Coal, oil, and natural gas are from prehistoric plants and animals that were nourished by sunlight, and winds and tides are from thermal changes due to the sun. Perhaps only geothermal and nuclear energy are exceptions. Direct capture of sunlight has been a dream for at least a century and a half. Nature's most efficient energy harvester, the green leaf, works using a series of chemical nanostructures that capture the sunlight, pool it into chemical reaction centers, and use it to make energy-rich chemicals and biomass while turning carbon dioxide into oxygen. Nanoscience approaches the problem in various ways, ranging from attempts to imitate leaves (usually called biomimetics) to more pedestrian approaches including highly porous structures in which sunlight is used to heat water and produce very high energy content steam.

None of these methods yet has a total efficiency much above 25% (though this is up from 3% just a few years ago), but through continued engineering and development we can anticipate that large-scale solar capture will become real and efficient. Startups like Nanosolar are already making better nanoenhanced solar collectors (Figure 5.4). Not only are these new cells more efficient, but Nanosolar claims that they are flexible, so they can be wrapped over any surface. Moreover, they are only 300 nanometers thick—around 1000 times thinner than the average silicon-based solar cell in use today.

Nanotechnology for energy conservation is also an area of active research. One promising approach works very differ-

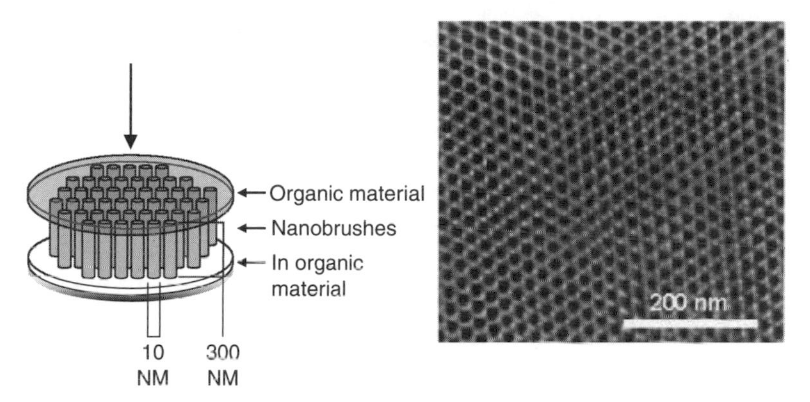

Figure 5.4 Nanoarchitecture image of an enhanced, flexible nanoscale solar energy collector. Courtesy of Nanosolar, Inc.

ently than most conservation methods because it doesn't target usage in businesses or at home. Instead, it targets the energy distribution network—the high voltage wires that transmit energy from generation plants to where it is actually used. These wires waste a vast amount of energy—estimates range around 30%—simply because copper wire cannot conduct electricity perfectly. Some of the energy is always dissipated as waste heat and the more distance the electricity travels, the more is wasted. Nanotubes offer an intriguing alternative since they can conduct electricity without resistance when properly configured. This means that a nanotube power cable would dissipate little or no power and immediately result in a remarkable reduction in energy needs without altering productive usage. Unfortunately, it will not be practical to make enough nanotubes for such a cable until fabrication technology for nanotubes gets much better.

With improvements like efficient solar power and nearly resistance-free transmission, the world can deal with climate change and global warming, and do so in an atmosphere of energy security. Adequate supplies of energy are needed if society is to permit its members to live lives of comfort and capability, without deleterious environmental impacts. The fact that energy production in the developing world is not yet sufficient to serve its nations' needs is a double-edged sword. On the one hand, it is certainly desirable that these countries should have the energy they need to have good living and working standards. On the other hand, if every country on earth generated energy in the environmentally harmful way that the developed world, notably the U.S., now does, climate change effects would be dreadful. Energy remains a global challenge and one of the great problems that we must and will solve.

Solutions like efficient solar power and clean nuclear fusion power may still be far away, but nanotechnology already impacts some aspects of the energy supply situation in an important way. For example, gasoline is made by "cracking" natural petroleum. Cracking is a chemical process in which large molecules in petroleum are broken down into smaller molecules, whose octane rating is high enough to be useful as a motor fuel. One of the major advances in the cracking process occurred when scientists at Mobil developed a series of catalysts for making higher-octane gasoline from the same petroleum feed-stock. The trick involved using minerals called zeolites, materials made familiar by their use in water softeners. Zeolites are mixed oxides of aluminum and silicon. They have nanoscale pores and chambers throughout their structures that make them look and act like molecular sieves. Cracking using particular zeolites, with appropriate pore sizes and structures, produces much higher octane gasolines. This improvement in making gasoline from petroleum is entirely a matter of nanotechnology, because the engineering of the pore sizes and pore crossings to make the best cracking catalyst involves an understanding of the structure at the nanoscale.

The contribution of these catalysts to the U.S. economy is billions of dollars per year already. Nanoscience's applications in the energy field have begun.

The Hydrogen Economy

Hydrogen gas is in many ways an ideal fuel. It has very high energy and very high power per pound. It is also a clean fuel, since it burns cleanly with oxygen and produces only pure water as a waste product. Hydrogen can be produced directly from water using electricity or possibly using sunlight directly.

Because of these striking advantages, there has been extensive discussion of using hydrogen as the entire basis of energy in our economy—to create a hydrogen economy rather than a coal or oil economy. This hydrogen-based energy model has a great deal to recommend it, including political and security advantages (there is more water in the United States than oil in the Middle East) in addition to the obvious environmental benefits.

In order for the hydrogen economy to be practical, we must be able to use, make, distribute, and store hydrogen efficiently. Some of these challenges are easier than others. Using hydrogen is the easy part. We've already looked at fuel cells that can burn hydrogen, and it can also be burned in more conventional turbines. Distributing it is also technically straightforward since it can be sent through pipelines as a gas or placed in high-pressure vessels (tank cars or cylinders) and moved by truck, train, or ship. Unfortunately, refitting existing systems may be quite expensive.

Producing hydrogen is a bit more complicated. It is best made from water since two of the three atoms in a water molecule are hydrogen. While it can be created from water by electricity in a process called electrolysis, the electricity required for this method has to come from somewhere, and therefore this process largely defeats the purpose unless the electricity comes from another clean source like wind, geothermal energy, or, in the more distant future, nuclear fusion. Other methods for extracting hydrogen are being examined, and intense research is now being devoted to the hard problem of using nanoscale methods for making hydrogen from water by direct photo conversion of sunlight.

The hardest nut to crack may well be hydrogen storage. Hydrogen is a gas at ordinary pressure and temperature, and it is very flammable and volatile (the Hindenburg airship disaster arose from the combustion of the hydrogen used to keep the airship afloat). While hydrogen can be stored at high pressure in tanks, the tanks are heavy, and it requires a good deal of energy to compress the gas. In most proposed hydrogen storage methods, the storage medium actually weighs a great deal more than the material being stored, making storage of hydrogen quite inefficient on a weight basis. Nanoscience, including biological nanoscience, offers a few possibilities, including storage in carbon powder, nanotubes, or some promising biomaterials. Solving this problem would be a major economic advance.

Domestic Economic Implications

Information tech, biotech, and nanotech all function as technological enablers. They encourage new investment, they produce new goods and services, and they provide new jobs. Because of the widespread applicability of nanoprocesses and nanostructures, nanotechnology could engender an economic boom similar to the one provided by information technology in the 1990s. Even after the first boom it will remain a massive force in the economy.

We've spent most of the book looking at security, defense, and energy applications of nanoscience. It's also worth taking a quick look at some of the near-term civil applications since these make up a large part of the economic picture. To take one simple example, roughly $45 billion per year is spent in the United States on medicines that have a problem with bioavailability. Drugs can be injected, ingested, or even breathed into the body, but it is difficult to make sure that the active ingredient in the drug gets to the part of the body where it is needed (a tumor or inflamed tissue for example), since most drugs aren't very soluble in water or blood. Most of the medicine actually passes right through the body without being used. The amount that gets to where it is needed compared to the amount consumed represents the bioavailability of a drug. With drugs as expensive as AIDS treatment cocktails (which can run over $10,000 per year), even a small amount of waste gives a major increase in cost.

While many approaches to the bioavailability problem seem reasonable, one of the simplest involves using the nanoscale. Because the rate at which a drug dissolves depends on the surface area of the powder or tablet, using grinding techniques to reduce the powder size to the nanoscale will substantially increase solubility and, therefore, bioavailability. Figure 5.5 demonstrates that simply grinding the medicine to reduce individual grains to below 300 nanometers in size results in much faster dissolution in water.

The effectiveness of the drug increased by several times, simply by grinding the particles down to the nanoscale. How much impact this new technique will have in solving the bioavailability problem is still unclear, but it is clear that the control offered by nanoscale behavior, even in something so simple as solubility, can have major implications in medicine

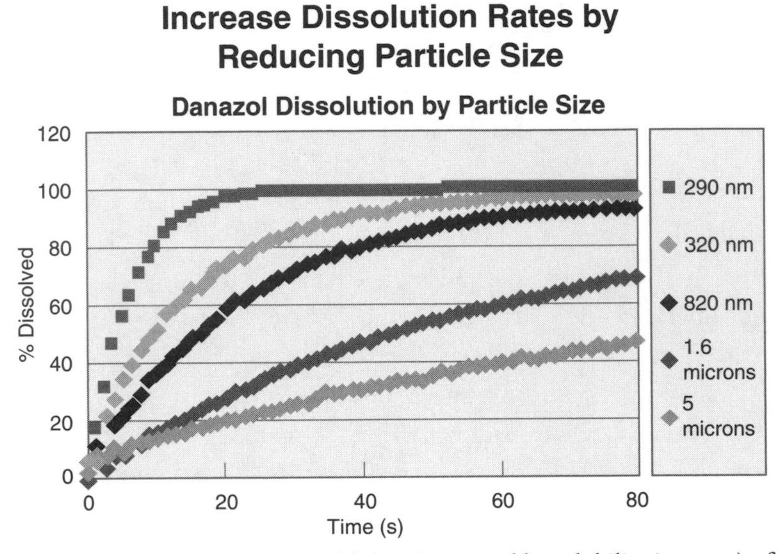

Figure 5.5 Increasing bioavailability (measured by solubility in water) of a medicine upon grinding it to nanoscale size. Courtesy of Chris Tucker, Dow Chemical Company.

just as it does in manufacturing, transportation, and other important sectors of the economy.

Awareness of the economic possibilities offered by nanoscience is widespread, and is one reason why *Spiderman* and the *X-files* kept *nano* in their scripts. But less televised developments also show how big it will be to get small. Phillip Bond, Undersecretary of Commerce, is bullish on nanoscience, saying, “It is phenomenal what the future could be in this space. It’s almost literally beyond most people’s imagination. It’s going to change everything.” The number of nanotech startups doubled to more than 1200 in the past year, and is expected to double again. By some estimates more than 5% of all the venture capital investments made in the United States in early 2003 went into nanotech or related fields. Mike Roco,

Director of the National Nanotechnology Initiative, has stated that in early 2003 more than 1700 companies in 34 countries were pursuing the commercial possibilities of nanotechnology.

More extended discussions of the possible implications of nanosciences on the economy can be found in other places, but it is good to remember that these are all estimates. Since we don't yet know how much investment there will be in the nanosciences or how rapidly the products of nanotechnology will be absorbed into the economy, these predictions are difficult. Nevertheless, a good number to remember is the estimate from the National Science Foundation that by the year 2015, nanotechnology could be responsible for GNP increases of roughly $1–2 trillion dollars per year. This number corresponds to about $6,000 per person per year in the United States and is representative of the kinds of advantages that nanoscience might provide for economic security.

Global Economic Impact

Nanotechnology is a so called "disruptive" technology. This means that it will have wide-ranging implications not just for improving current technologies, but also for creating new kinds of devices and capabilities with ramifications for the very way we live our lives. Previous examples of disruptive technologies include biotechnology, information technology, flight, and electricity. Nanotechnology may be as transforming as any of these.

There is another ramification of many disruptive technologies. Since they are not necessarily built on the foundations of older technologies, they provide opportunities for anyone to get ahead. In Chapter 3 we briefly discussed bioelectronic computing. The story behind this is an interesting illustration of why disruptive technology is so important. Research in bioelectronic computing really began in earnest during the 1980s. Leaders in Russia (still the Soviet Union at that time) feared the United States' dominance in the then emerging field of information technology, and they worried about how to compete head to head in developing the new technology. Instead of putting all of its resources into improving conventional computing, Russia therefore invested heavily in research to develop totally new computing techniques not based on silicon. If it could be the first to develop these, and if their potential outstripped that of silicon electronics (or their manufacture turned out to be easier), the Soviet Union could leapfrog the United States in the information race. The plan didn't work out quickly enough for the Soviet Union—bioelectronics and other architectures proved evasive and are only now becoming reality—but the initiative was strategically sound. Disruptive technology gives little or no advantage to the entrenched leader of an earlier technological wave.

Another example of the power of disruptive technologies is India's leadership in the software sector. India did not start with any natural advantages in developing software when it starting trying to create a software industry in the 1980s and 1990s. It did not have a strong semiconductor business, computers were difficult to import, its telecommunications infrastructure was primitive, and almost all of the basic technology was being created overseas. However, by combining extensive private and public investing with crucial government reforms and strong entrepreneurial leadership, India has outstripped the industrialized economies of Europe and now produces around a third of the world's software. Software has become a key sector of the entire Indian economy and, from an international perspective, probably the most important sector of all.

Nanotechnology could work the same way. China, Japan, Australia, Canada, and most European nations have already set up national nanotechnology initiatives of their own. The governments of Russia, Israel, and other powerful nations have also established national nanotechnology centers. While the budgets vary widely, some of these efforts (notably China's and Japan's) are comparable in size to America's, and some of the best work in nanotechnology is coming from overseas. Since all of these nations have excellent scientists generating world-class work, the question of which economies (and militaries) will be most affected by nanotechnology is still very much open to contention. The answer to this could have a great effect on the balance of power.

The Economics of Security

Economic health and national stability go hand in hand. The poorest nations in the world are also the most troubled—North Korea, most countries of Africa, and most of the nations of the Middle East are prime examples. Economic downturn has led to coups, riots, and even the collapse of the Soviet Union. By its disruptive nature, nanotechnology provides a new set of opportunities for those who embrace it. The potential for it to be an industry generating $6,000 in GNP for every citizen of the United States within 15 years is particularly striking when you consider that this increase alone is much more than the total yearly income of most of the world's people. Many of these will now have the opportunity to use nanotechnology to follow India's software example.

Another important security aspect of nanotechnology lies in its global nature. Thomas Friedman, the *New York Times*'s foreign affairs columnist, has proposed that the tighter economic ties between countries involved in global trade and commerce have resulted in a "golden straitjacket." This straitjacket is a means of gaining great wealth through trade, but it carries with it several constraints including the requirements that trade must remain open and international relations must remain warm. War between two nations with free market economies and truly democratic governments is so rare as to be almost inconceivable. If it continues along its current development path, nanotechnology will continue to grow internationally and become one of the great engines of international commerce. If Friedman is correct, that alone would be a major contribution to international security.

Society, Ethics, and Geopolitics

"Where there is no vision the people perish" —*Proverbs 29:18*

> Stan Williams
> Senior HP Fellow and Director of Quantum Science Research
> Hewlett-Packard
>
> We are the descendants of people who successfully embraced a series of revolutionary new technologies and the changes they enabled. Succeeding generations of our ancestors used their new tools and materials to construct and protect their civilizations. Nanotechnology is our generation's great opportunity and responsibility. Properly developed, nanotechnology has the potential to improve every human-made product and enable entirely new objects not yet conceived. In the shorter term, nanotechnology offers the best defense agains the malicious use of previous technologies.

The last chapter of this book differs from the others because it is much more subjective. It deals with policy, society, and ethics, which are not matters of science or technology alone. These views cannot be proven like mathematical theorems, but we've chosen to include them because we think it is important to present a balanced picture, to show some of the potential problems as well as the benefits of the technology, and to start engaging these issues as soon as possible.

Concerns about technological and scientific advances are not new. In the early 19th Century, the Luddites were concerned about the introduction of machinery that would displace workers, so they destroyed the looms in their workplaces. Eminent scientists such as Lord Rayleigh didn't believe in human flight, saying in effect that if the Lord wanted us to fly He would have given us wings. More recently, genetic cloning and the use of stem cells in medical research have been front-page news, provoking not just debate within the scientific community, but sermons, demonstrations, and even riots. The Luddites' name has become synonymous with the shortsighted opposition to any kind of technology, and the remark about flying is now proverbial for its lack of vision, but there are real issues to discuss regarding the emotional reactions to nanotechnology. Nanotechnology represents a wholly new set of possibilities for human advancement, so a thoughtful and reasoned policy comprised of legal, educational, and financial decisions is needed to allow people worldwide to maximize its benefits and control its risks.

Information Tech, Biotech, Nanotech

Within recent memory, two other major disruptive technologies have been developed: information technology and biotechnology. Examining their records reveals two paths for how nanotechnology could develop.

Information technology has, for the most part, been a good citizen. Since computers, networks, and the Internet have become common, industry groups like the Internet Engineering Task Force (IETF)and the Institute of Electrical and Electronic Engineers (IEEE) have done a good job of promoting standards of interoperability. Schools and professional training programs have smoothly (or by fits and starts) integrated computers into their curricula. Periodicals and training manuals exist for every level of expertise from the beginner to the expert, and industry watchdog groups such as the Electronic Frontier Foundation (EFF) and the Center for Democracy and Technology (CDT) watch both government legislation and industry practice for illegal or unethical activity. Other groups, including the Computer Emergency Response Team (CERT) and independent computer security vendors, track the spread of computer viruses and software vulnerabilities with such efficiency that major newspapers carry notifications of big problems and fixes are usually released within days. Despite recent antitrust issues, the industry remains quite open and competitive and, for the most part, it has been remarkably constrained in its lobbying for legislative protection (though the same cannot be said of the telecommunications industry). Many questionable laws such as the Communications Decency Act have been defeated or repealed, and potentially threatening legislation like the Total Information Awareness Act are openly debated. Intellectual property protection remains strong enough for companies to make respectable profits, but solid alternatives for almost every software product now exist within the open source community, many of them in the public domain.

Biotech's record is spottier. While basic biology is part of primary education, most people don't have a good understanding of the issues involved. There is less engagement by the press and there are fewer journals for nonexperts. Also, since there is less concern for how different biotech applications interoperate than there is for computer parts, there are fewer open industry associations setting standards. The level of secrecy in biotech research largely prevents watchdog groups from being effective, but this is to some extent justified since the biotech product development cycle is so much longer than it is for information technology (years as opposed to months), and a key scientist leaving a project can set it back almost to its beginning.

The stakes are also higher in biotech than they are with information technology. Few people march or riot over new software releases (tempting though it can sometimes be), but the biotech industry is almost always vilified as meddling with nature or playing God, even though it has generated greater advancements in medical treatments and in prolonging life than any other technological revolution in history. The stakes are also higher because when research is done incorrectly or when companies act unethically, their behavior is likely to result in serious side effects or death, often at a massive scale. Biotech is an industry that needs more ethical scrutiny than it has received.

Nanotech sits at the crossroads between scientific and engineering disciplines, and it has a lot to learn from each. We hope it will be able to combine the best of both, and we now turn our attention to particular areas.

Public Policy

People may have started, as Rousseau thought, in the state of nature. But the nation state is the organizing structure of current societies almost everywhere on earth, and it is within these nation states that nanotechnology will emerge. People cannot plan for nanotechnology as individuals, so it will be necessary for public organizations to embrace the future and plan for it.

In the United States, the National Nanotechnology Initiative (NNI) was launched during the Clinton administration. This initiative is now funded at a level of almost $1 billion per year, with much of this funding going into research and development efforts at universities, national laboratories, and industrial research operations. NNI also has important educational, policy, ethical, and outreach components. Some regulatory structures that were put in place for other concerns, including everything from zoning laws and air pollution regulations to noise abatement statutes and building codes, deal with aspects of nanotechnology. Government-supported education and training initiatives and development zones (including nanotechnology incubators) are also responses to nano. For these reasons NNI, plus extant regulatory structures, represent the nucleus of current nano policy.

More is needed. In Chapter 5, we discussed the environment, energy, and economics. Since the economic impact of nanoscience will be very large, it is the responsibility of public policy to anticipate its effects and to try to cushion some of the possible negative ones. This includes planning for the economic shock that may come if competitive new products and nanomanufacturing cause job losses in older manufacturing sectors. Public policy must also provide appropriate new regulatory bodies to handle the unique challenges of nanoscience.

The set of challenges that nanotechnology will provide might seem intimidating. It is clearly wise for government to look before it leaps, but inaction or unending delays could be destructive. The ETC Group, an international public interest body, advocates a total moratorium on nanotechnology until all of its environmental, economic, and other impacts can be evaluated. This approach is not feasible. Nanoscience is already here and nanotechnology is arriving. Both are already providing great benefits for the society and will provide many more. Delay is not feasible (and from a defense perspective it is strategically suicidal, for even if the work is not done in the United States, it will certainly continue elsewhere), but regulation, planning, and education are needed.

Watchdog groups like ETC are important contributors to the debate. Their studies, along with academic review and open research policies by institutions doing nanoscience research, provide crucial transparency similar to that enjoyed by information technology. Only with this kind of transparency can the nanoscience community hope to maintain public trust while pursuing research that many people find threatening. In addition, it is important to keep public attention focused on nanotechnology. Continued press coverage, open industry meetings, and even coverage in popular fiction are important components of this. On the government side, specialized task forces, panels, committees, reports, and legislative and executive actions help keep attention focused, and they allow for speedy decisions on nanotechnology policy when necessary. This speed will be essential as the nanotechnology industry continues to gather momentum. The rate of innovation for nanotechnology could well exceed that of either information tech or biotech, and the decisions required by the government to regulate it could be at least as complex.

Why is this so important? The discussion surrounding genetically modified (GM) food provides a prime example. In 2002 the United States promised famine relief to Africa, shipping tens of thousands of tons of grains including genetically modified corn. Despite the threat of starvation, African governments in Mozambique, Zambia, and Zimbabwe hemmed and hawed about whether to accept the food, saying they were unsure if it was safe to eat, plant, or even feed to livestock. Genetically modified food of this type has been common in America for years, but regulators in Europe had escalated fears about the safety of GM food, largely due to pressure from farmers' lobbies. Even after years of using GM crops in the United States, old-fashioned trade wrangling combined with the GM food industry's perceived lack of forthrightness in providing impartial studies about the safety of their products resulted in this crisis. It had nothing to do with science or technology, only with irresponsible policies and a lack of education. Governments have a responsibility to prevent this kind of thing from happening.

Intellectual Property

Technologies are developed by people and organizations, and these inventors have the right to protect their discoveries and developments by securing patents. Nearly all developed countries have a patent system, and some larger entities such as the European Union and World Trade Organization (WTO) are developing even broader patent policies. Patents will be as crucial for nanotechnology as they have been for other technologies. Inventors, developers, and investors need to protect their ideas, devices, and investments. Reasonable patent protection is a necessity for nanotechnology to flourish and to contribute its unique attributes to all aspects of society, including homeland security.

But a key difficulty lurks in the word *reasonable*. For an example, far from nano anything, think for a moment about children's literature. In 1998 Congress approved an extension of copyrights often called the "Mickey Mouse Law," since it was passed under enormous pressure from Disney just before Mickey went out of copyright. In Mickey's case the law extended the copyright to a total of 95 years after first publication. Mickey first appeared in 1928, so this extension to the copyright will preclude any other creators from using Mickey (or Goofy and Unca Scrooge, either of whom might have provided a better name for this copyright policy) until at least 2023. This hyperextended copyright protection reaches well beyond any reasonable lifespan of the creators and seems arbitrary, unreasonable, and suffocating.

J. K. Rowling, now richer than the Queen of England and worth some $500 million, is seeking very broad copyright protection for her character Harry Potter. She and her publisher are now using copyright protection to limit other creative work involving child magicians. For example, they have sought to block the western publication of a story about a Russian girl apprentice magician called Tanya Grotter as well as Chinese and Bengali take-offs on Rowling's idea. A case could be made that these are noninfringing "fair use" creations covered under laws designed to protect parodies, or that they are derivative works that do not deprive an author of reasonable profit from her work. By pursuing an attack on these works, Rowling and company are potentially stifling other authors whose work could in no way be mistaken for the original Harry Potter stories but could delight children who are culturally removed from Harry (British-style boarding schools, myths, and values not being universal). Based on this behavior, their view of copyright protection seems as draconian as anything Voldemort might have conjured up.

How does this apply to nanotechnology? Copyrights and patents are intricately related, and intellectual property law will be key to the development and use of nanotechnology. In the pharmaceuticals area, for example, patent issues are particularly complex since they can have life and death consequences. Without patent protection, big pharma cannot justify the very large expenditures involved in drug development; it can cost more than $100 million to develop, test, and certify a drug. However, protection that is too strong can be abused, and serious abuses have already been committed. Abuses can take a variety of forms. A few common ones include patenting a drug but not producing it (either to protect a competing profitable drug or to await a more lucrative offer for distribution), holding a patent back while demanding an exorbitant licensing fee, and keeping prices artificially high long after development costs have been recouped.

Patent law is already very broad. It is possible to patent individual genes, for example. This means that a patent-holder can choose to delay the development of therapies and treatments for any reason at all. Not reading Harry Potter in a Russian setting may not work significant harm to society, but a 20-year delay on a drug for halting breast cancer certainly would. While the pharmaceutical industry has produced remarkable drugs that have made our lives richer, healthier, and longer, the ethical issues that surround patent policies are vexing.

One other aspect of globalization is that it now takes much less time than it has at any other point in history for a product to reach world markets once it is approved for release. Harry Potter was released in markets worldwide on the same day, and drugs can be on pharmacy shelves across the country within weeks after approval by the Food and Drug Administration (FDA). The World Trade Organization has opened many new markets and is working to provide equal protections in all of them. All of this means that it should be easier than ever for companies (or authors) to profit from their inventions in less time than ever. This would seem to make the case for shortening rather than extending protection, and there should be some review of when public policy considerations should allow specific intellectual property rights to be overridden. Led by Brazil, which ignored some international patents to create one of the world's most effective and affordable AIDS treatment programs, this is indeed happening in the developing world. Even as authors, we think it would be nice to live in a world where we could sing iconic American songs by George Gershwin or Irving Berlin without having to slip a copyright notice into the chorus.

In nanotechnology, the patent issues have not yet captured public attention, but clearly there is an overwhelming human need to see that the technology is developed and used. Patents should advance both the society's progress and the developer's income, and we do not want the situation to be similar to the worst of Harry and Mickey on drugs.

Privacy and Civil Liberties

The Constitution of the United States does not specifically mention the right of privacy, but it is a remarkable, strong charter protecting civil liberties. Issues of privacy and civil liberty within American society are clearly much larger than any simple nanotechnology considerations, but nanotechnology does add urgency to them. Our progress toward the use of nanostructures within the society should engender a new concern about secure protection of privacy and civil liberties.

Shakespeare's Brutus was concerned about Caesar taking too much power when he said, "And then, I grant, we put a sting in him, / That at his will he may do danger with." There are obvious ways in which nanotechnology will substantially increase the capability of governmental and large private organizations to ride roughshod over privacy and civil liberties. For example, if we develop the capability to screen any individual's genome in one day for a thousand dollars (and this will happen before the end of this decade), then whose property is that information? The genome will contain, for example, indicators about inherited diseases, allergies, different medical dysfunctions, and genetic traits. Already, screening for particular markers (like the brac gene for breast cancer) has been a great advance in preventive medicine. Full knowledge of a genome would be a great medical advance for the individual patient, whose awareness of his or her own condition would be very much improved. But if these records become the property of employers, insurance companies, or government agencies (or boyfriends, or girlfriends, or mortgage brokers) then the individual's rights have been invaded.

In a total information society where nano-enabled pervasive computing is present everyplace, the ability to protect citizens' privacy is challenged. Things like grades in school, memberships in clubs and organizations, sexual preferences, consumption patterns, and political views can all be captured. Nano-based code breaking would not only allow better cracking of terrorist plots, it could also prevent any personal computer files or Internet browsing habits from remaining private. This could be both a government problem and a corporate one. The Recording Industry Association of America has advocated and received some legislative support for broad rights allowing it to defend its copyrights aggressively by breaking into computers of those it suspects of sharing music files on the Internet. If support for these policies continues and if advanced cryptographic techniques become available, the music industry (and any other similarly treated industry) could easily create a database of personal information more impressive than the FBI's.

To whom does this information belong? To whom should it be revealed? What are the constraints on disseminating that knowledge? These issues, again, predate nanoscience. But when we can monitor individual conversations through walls and windows, when old-fashioned wire taps are replaced by electromagnetic monitoring, and when nanotechnology allows labeling and tracing of every individual pill, computer chip, and pizza box, then the possibilities for loss of privacy and civil liberties are more frightening. At a time when the Attorney General of the United States is pursuing ever broader surveillance rights and continues to develop plans to reward the citizens for spying on one another these privacy concerns are immediate and deep. Clearly, effective regulation arising from awareness of capabilities and the importance of privacy is part of the answer. Nanoscience simply creates a more pressing need for society to address these issues.

It would seem consistent with the complexity of our society and with many Supreme Court findings to assume an implied federal privacy right. However, the capabilities of nanotechnology, biotechnology, and information technology demand further and more permanent protection. Within an American context, it seems that only a Constitutional amendment directly stating a right to privacy can be trusted in light both of inconsistencies of government policy (as can be seen in the USA "Patriot" Act and the proposed Total Information Awareness Act) and of these advances in technology.

Education and Training

The invention of the automobile required driver's education. The invention of the aircraft required pilot training courses. The development of information technology required many of us to learn more than we might want to know about operating systems, file formats, and networking components. The invention of the telephone brought telephone operators, linemen, telemarketers, and AT&T. Similarly, the development of nanotechnology will mean new jobs, new careers, and new work patterns. It will also mean displacements, economic disruption, and changing employment and investment patterns. Nanotechnology should represent an opportunity for economic gain and for better peace prospects worldwide.

To enable society at large and individuals within that society to take maximum advantage of the opportunities and to overcome the challenges of nanotechnology, new educational systems will be required. With appropriate education and training, nanotechnology and its products will be no scarier than the bicycle, which was also a tremendous technological advance when it was developed. Education and training can provide security, employment, understanding, and ease in the society as it will exist after nanotechnology has added its modicum of new things. Even in *Prey*, his novel on nano disasters, Michael Crichton shows that the perceived nano dangers can be solved by thought and by education.

People's fear of the unknown can be palpable, and can lead to riots, ostracism, xenophobia, government change, and war. Education, training, and openness are ways around these disruptions. Globalization, economic change, and rapid innovation become familiar, rather than frightening, when our education permits us to understand what they are about.

Nanotechnology will provide new medicine, new energy, new materials, new fabrics, new tennis rackets, new deodorants, new cheeses, and new skis. We will be able to see, hear, smell, and taste better. Since some aspects of this technology will be strange, it will be the responsibility of our educational systems—public, private, and supplemental—to help us understand these things. Books, Web pages, articles, lectures, newspaper columns, and television specials will help us realize the huge capabilities that nanotechnology brings.

Amorality of Technology

Science and technology are at their base value-free and amoral. Insight and wisdom can be used positively or negatively, depending on education, moral values, and social issues. In the words of Deuteronomy 30:19, "I have set before you life and death, blessing and cursing: therefore choose life, that both thou and thy seed may live."

All of science—indeed all of human effort—empowers society in all directions, good and bad. Radio brought communications and safety, but it also brought the boom-box and electric instruments and noise pollution. Prometheus the fire-giver is an image of power, but fire can burn destructively. The Haber process that makes synthetic ammonia from nitrogen and hydrogen was developed at the beginning of the 20th Century, and the fertilizers it produces are essential for world agriculture to feed the billions of people on earth. But the Haber process also makes it much easier to make artificial gunpowder, a result that helped extend World War I, killing millions in the process. In all of these examples, science and technology enable society to do critically important things both good and bad, and it is the responsibility of the society to decide how to use new ideas. Education will help, and all the moral and societal values that we develop will help determine how the world responds to the capabilities of nanotechnology.

Henry David Thoreau claimed that "time is but the stream I go a-fishing in." But the current in the stream has sped up. We now measure time in nanoseconds or less, and as this book has tried to make clear, structures at the nanoscale determine behavior in our tangible and real world. The total information society will make everybody more aware, simply because more information will be available to everybody.

The promises and threats from amoral science and technology are greater than ever. Global society will change tremendously in this century, and nanoscience will play a significant part in this change.

Vigilance, Awareness, and Responsibility

In the Constitution for the State of Massachusetts, John Adams required that the government of the State further the arts, the sciences, and education as part of its mission. But the individual members of our present society must force the establishment of appropriate government regulatory structures based on knowledge, insight, tradition, and individual rights.

Admiral David Jeremiah, former vice chairman of the Joint Chiefs of Staff, understood the need for vigilance in a world in which the MAD doctrine is no longer adequate when he said, "Military applications of molecular manufacturing have even greater potential than nuclear weapons to radically change the balance of power. In anticipation of that possibility, the uninformed policy maker is likely to impose restrictions on the development of technology in such a way as to inhibit commercial development (ultimately beneficial to mankind) while permitting those operating outside the restrictive bounds to gain irrevocable advantage."

The need, then, is for vigilance, awareness, and action on the part of society. We no longer live in small villages isolated from one another, but in a huge, global village. We can no longer just move to avoid the smoke from the neighbor's chimney, as Abraham Lincoln's family did. Instead, we need to embrace the capabilities and possibilities that technology offers, and to do so within a societal structure that is functional, creative, responsible, and protective of individual rights.

In an article from *Defense Horizons* in March of 2002, Peterson and Egan make the following argument about instability brewing in the world.

> *[Instability is found in] those places where poverty, lack of education and lack of human rights are concentrated. In fact, one could argue that the dichotomy between the haves and have nots (both economic and digital) is by far the greatest looming global security issue ... For the first time in history, a new technology holds forth the promise of providing*

inexpensive food, energy, clean water and probably education for everyone on the planet. Nanotechnology could also be used in innovative ways to encourage national political stability and responsibility. We should begin to think about the future in these terms, for we have a choice: either we'll be defensive and respond to problems as they arise, or we will shift to the offensive and use the military and these new tools in creative new ways to deal with the problems while still we can.

Regulation is one of the legitimate activities of government at all levels. The existence and enforcement of local, county, and state regulations such as noise abatement statutes, air and water quality standards, workplace safety regulations, zoning laws, and toxic waste disposal requirements are very important. These constitute the first level of society's defense in light of the environmental and safety issues associated with nanotechnology and all emerging technologies.

More is needed. Federal action on such issues has become a political football. The current Bush administration has both pluses (the initiatives on hydrogen fuel cells, nuclear fusion, and support for the NNI) and major minuses (refusing to deal responsibly with global warming, advocating drilling in the Alaskan National Wildlife Refuge, and alteration and lagging enforcement of environmental standards). The issues that advanced technologies present require federal attention in both the development and regulatory sectors. We need an FNA (Federal Nanotechnology Agency) to complement the tremendous development efforts of the NNI with appropriate regulation, policy advice, product approval, and monitoring. This agency should undertake development, adoption, and enforcement of statutory and regulatory aspects of nanotechnology and associated advanced technologies.

Just as flight brought the FAA, as radio and TV brought the FCC, and as food health concerns and drug development brought the FDA, so nanotechnology requires (and society deserves) an independent FNA with deep expertise and broad powers. The reason to propose yet another federal agency, rather than to assign responsibility to existing alphabetical organizations, is that nanotechnology is so different. For example, according to ETC Group reports, the FDA has assigned so-called "functional equivalence" (a designation indicating that a new food or drug operates on the same premise as another and therefore requires less rigorous approval) to nanoscaled structures, such as titanium dioxide nanoparticles used in skin cream and suntan lotion. This means that their toxicity levels may receive less attention than is due, and major concerns over the carcinogenic properties may be glossed over without adequate testing. This reflects a lack of understanding of the vast differences at the nanoscale.

Just as globalization is a major theme in nanotechnology's development and application, so global approaches are needed for effective regulation. As an example of what does work, there is the Montreal Agreement that led to the protection of the ozone layer by phasing out worldwide production of chlorofluorocarbons (CFCs). An example of what doesn't work is the failure to date of the Rio/Kyoto approach to the control of global warming. We now have the UN, the WTO, the World Health Organization (WHO), and the World Bank. We need a World Nanotechnology Organization (WNO), perhaps as a working group within one of these larger and established organizations. The WNO structure must be inclusive—advanced and developing countries, government and labor, industry, public interest, and academic representatives must all be included. To operate, the WNO will require open meetings and clear mandates and powers. This kind of inclusive policy may account for the success of the Montreal Agreement, just as the failure to date of Rio/Kyoto may be due to its lack of inclusion. This WNO should both regulate and document the technology. It should set standards and help in their enforcement. It should enlist the best minds and the best efforts of humanity in the control, understanding, and applications of nanotechnology.

Visionaries from Isaiah and Zamenhof (the inventor of Esperanto) to Woodrow Wilson and Grotius (founder of international law) have envisioned both world peace and world security through understanding, law, morality, and policy. New capabilities for great economic security through energy, science, biotechnology, nanotechnology, and information technology can hasten the betterment of the world, and of its individual citizens.

The answers to these challenges lie in action and in education, in understanding and in responsibility, in civic awareness and in reaffirmed respect for privacy, civil rights, and individual determination. Big Brother is not the answer. We need the collective will and wisdom of all peoples. We need vision, humility, and peace. Nanoscience and nanotechnology will lead to wealth, to security, to better health and better living, but they will only do so if society welcomes them with awareness, education, and responsibility.

Index

A

B

C

D

E

I

J

K

L

M

O

P

Q

R

S

X

Z

About the Authors

Daniel Ratner is a veteran of high-tech startups, currently serving as Executive Vice President and CTO of Driveitaway.com, the first automotive industry–specific Web auction. Prior to Driveitaway.com, Mr. Ratner was the co-founder and CTO of Wired Business, one of the first nationwide providers of DSL Internet access. He began his startup career as the founder and CEO of Snapdragon Technologies, an IT consulting firm specializing in information systems and strategies.

Mr. Ratner holds a B.A. degree in engineering and economics from Brown University and he is a Visiting Scholar at Northwestern University, where he has lectured on nanotechnology at the Kellogg School of Business. He sits on the Board of Directors of Sittercity Inc. and The Chicago Society of Music and on the Board of Advisers of First Colonial National Bank and RMS Investment Corporation. He has also been a mentor for the Brown University Entrepreneurship Program and a speaker on nanotechnology at many business conferences and trade events. In July 2001 Mr. Ratner was selected by *PhillyTech* magazine as one of the "Thirty Under 30" entrepreneurs to watch in the Philadelphia area. With Mark Ratner he published *Nanotechnology: A Gentle Introduction to the Next Big Idea* (Prentice Hall, 2002).

Professor Mark A. Ratner is Morrison Professor of Chemistry and Associate Director of the Institute of Nanotechnology and Nanofabrication at Northwestern University. His lifelong work in molecular electronics, a field he is credited with creating in 1977, led to his receiving the 2001 Feynman Prize in Nanotechnology and becoming a member of both the National Academy of Sciences and the American Academy of Arts and Sciences. He has published four hundred scientific papers and two advanced textbooks on chemistry, nanotechnology, and related subjects. His first book for the general public was *Nanotechnology: A Gentle Introduction to the Next Big Idea* (with Dan Ratner).

Professor Ratner has been a named lecturer on molecular electronics and nanotechnology across the world, but he concentrates his efforts at Northwestern University (where he received his Ph.D., served as Associate Dean of the College of Arts and Sciences and Chair of the Chemistry Department, and received the Distinguished Teaching Award, appearing on the Faculty Teaching Honor Roll ten times). Professor Ratner holds a B.A. from Harvard University and he has held Fellowships from the A. P. Sloan Foundation, the Advanced Study Institute at Hebrew University, the American Physical Society, and AAAS.